Introduction to Astronomy: A Fun Exploration of the Universe

The Science of Astronomy

Astronomy is like a giant treasure hunt across the universe! It's the scientific study of stars, planets, comets, galaxies, and everything else beyond Earth's atmosphere. Imagine using a superpowered magnifying glass to look at the biggest and smallest things out there. From navigating ancient seas to launching modern space missions, astronomy has always been about discovery and adventure.

Astronomy splits into cool subfields, each with its own cosmic puzzles. Astrophysics uses physics to figure out how stars and galaxies work. Planetary science checks out the planets and moons, especially in our solar system. Cosmology dives into the universe's origins and fate, asking big questions like, "How did it all start?"

Educational Focus: This book will take you on a journey through basic and advanced astronomy. You'll learn about the tools astronomers use to explore the stars, the lifecycle of celestial bodies, and mind-blowing theories about the universe. Modern tech, from space probes to mega-telescopes, helps us learn more every day. By the end, you'll see how astronomy connects to physics, chemistry, and geology and how it helps us understand life on Earth and beyond.

Historical Perspectives in Astronomy

The history of astronomy is like a global detective story: Ancient civilizations saw celestial events as divine messages. They were the first to start recording the skies. The Babylonians, for example, kept impressive records of astronomical events.

In ancient Greece, astronomers like Hipparchus and Ptolemy built models of the solar system. They thought Earth was at the center, but their work set the stage for future discoveries. Then came the Renaissance and a game-changer: Copernicus proposed the Sun was at the center of the solar system.

The 17th century was like the blockbuster era of astronomy. Johannes Kepler discovered the laws of planetary motion. Galileo Galilei made groundbreaking telescope observations, and Isaac Newton's laws of motion and gravity unified our understanding of space and Earth.

Fast forward to the 20th century: Radio astronomy began, space exploration took off, and we got our first pictures of Earth from space. The Hubble Space Telescope, launched in 1990, has given us stunning images of distant galaxies and nebulae, expanding our view of the universe.

Did You Know? The Antikythera mechanism, an ancient Greek astronomical calculator, dates back to around 100 BC. It could predict astronomical positions and eclipses, showing just how advanced ancient civilizations were.

Modern Astronomy and Its Tools

Modern astronomy is a high-tech adventure! Telescopes and satellites are our windows to the universe. Galileo's telescope discoveries, like Jupiter's moons, changed our view of the cosmos.

Ground-Based Telescopes: Located in observatories worldwide, these telescopes use big mirrors or lenses to capture light from celestial objects. They can see different light wavelengths, from visible to radio waves. Examples include the Keck Observatory in Hawaii and the Very Large Telescope in Chile.

Space-Based Telescopes: These orbit above Earth's atmosphere, giving clear, undistorted views. The Hubble Space Telescope, for example, has been sending breathtaking images and data since 1990, revealing secrets about the universe's expansion and dark energy.

Radio Telescopes: Detect radio waves from space. They let astronomers study phenomena invisible to optical telescopes. The Arecibo Observatory, before its collapse, was a giant in radio astronomy.

Satellites and Space Probes: These extend our reach beyond the solar system. The Voyager missions, launched in 1977, ventured into interstellar space. The Curiosity rover on Mars is exploring the planet's geology and searching for signs of past life.

Theoretical Tools: Physics laws and computer models help us understand things we can't see directly, like black holes and dark matter. Simulations of the early universe test theories about cosmic structures' formation and evolution.

Educational Insight: As you color through this book, you'll see illustrations of these amazing tools and learn about their roles in modern astronomy. Discover how each one helps us explore the cosmos, from planetary surfaces to the farthest galaxies.

With this fun and colorful guide, you'll journey through the universe, learn about the stars and planets, and discover the incredible tools astronomers use to unlock the mysteries of space. So grab your crayons, and let's explore the cosmos together!

Chapter 1: The Solar System: A Cosmic Neighborhood
Chapter 1: The Solar System: A Cosmic Neighborhood

Welcome to the Solar System!
Welcome to the Solar System!

Welcome to our celestial neighborhood, where the Sun, eight diverse planets, and countless minor bodies like dwarf planets, asteroids, and comets reside. Each member of this cosmic family has its own unique characteristics and mysteries to explore.

Understanding Our Solar System
Understanding Our Solar System

At the heart of our solar system is the Sun, a massive star holding 99.86% of the system's mass. Around it, the planets orbit: Mercury, Venus, Earth, Mars, Jupiter, Saturn, Uranus, and Neptune. The shift from the ancient geocentric model (Earth at the center) to the heliocentric model (Sun at the center), initiated by Copernicus in the 16th century, revolutionized our understanding of the universe.

The Rocky Inner Planets
The Rocky Inner Planets

Closest to the Sun, the rocky inner planets have fascinating features:

Mercury:
- Extreme Temperatures: Swings from -180°C (-290°F) at night to 430°C (800°F) during the day.
- Surface Features: Craters, ridges, and volcanic terrains. Notable feature Caloris Basin.
- Thin Atmosphere: Composed of oxygen, sodium, hydrogen, helium, and potassium.
- Magnetic Field: Weak but present, about 1% as strong as Earth's.

Venus:
- Thick Atmosphere: Mainly carbon dioxide, creating a runaway greenhouse effect with surface temperatures around 465°C (869°F).
- Surface Conditions: Pressure 92 times that of Earth, like being 900 meters underwater.
- Volcanic Landscape: Volcanoes, plains, and mountain ranges.
- Retrograde Rotation: The Sun rises in the west and sets in the east.

Earth:
- Habitable Zone: Perfect distance from the Sun to maintain liquid water.
- Diverse Ecosystems: From lush forests to icy poles, teeming with life.
- Protective Atmosphere: Filters harmful solar radiation and regulates temperature.
- Dynamic Surface: Constantly changing due to tectonic activity and erosion.

Mars:
- Red Surface: Iron oxide-rich soil gives it a reddish appearance.
- Largest Volcano: Olympus Mons, 22 km high, three times the height of Mount Everest.
- Canyons and Valleys: Valles Marineris, a canyon system stretching over 4,000 km.
- Signs of Water: Evidence of ancient rivers, lakes, and minerals formed in water.
- Thin Atmosphere: Mostly carbon dioxide, contributing to its cold, dry environment.

The Gas Giants and Beyond

Further from the Sun, the gas giants and icy worlds dominate:

Jupiter:
- Largest Planet: Diameter of about 143,000 km, more than twice the mass of all other planets combined.
- Great Red Spot: A giant storm raging for at least 400 years.
- Moons: 79 known moons, including the Galilean moons. Ganymede is the largest moon in the solar system.
- Magnetic Field: The strongest of any planet, extending far into space.

Saturn:
- Ring System: Famous for its bright, extensive rings composed of ice and rock.
- Gas Giant: Diameter of about 120,000 km, primarily hydrogen and helium.
- Moons: 83 known moons, with Titan being the largest and having liquid methane lakes.
- Low Density: Less dense than water, it would float if placed in a large enough body of water.

Uranus:
- Tilted Axis: Rotates on its side with an axial tilt of 98 degrees.
- Ice Giant: Composed of water, ammonia, and methane ices. Diameter of about 51,000 km.
- Blue-Green Color: Due to methane in its atmosphere.
- Rings and Moons: 13 rings and 27 known moons, including Miranda with its extreme surface features.
- Moons: 14 known moons, with Triton showing geological activity.
- Blue Color: Methane in the atmosphere with an unknown component enhancing its vivid blue hue.

Neptune:

- Windiest Planet: Winds reaching up to 2,100 km/h.
- Great Dark Spot: A giant storm similar to Jupiter's Great Red Spot.
- Ice Giant: Diameter of about 49,500 km, similar composition to Uranus.
- Moons: 14 known moons, with Triton showing geological activity.
- Blue Color: Methane in the atmosphere with an unknown component enhancing its vivid blue hue.

The Sun: Source of Life and Light

The Sun anchors our solar system and fuels life on Earth with its energy. Its dynamic surface, with solar flares and sunspots, influences everything from satellite communications to the auroras. Engage with interactive activities to explore the Sun's surface and witness solar phenomena like eclipses.

Exploring Minor Celestial Bodies

Beyond the eight main planets, our solar system is teeming with smaller bodies:

- Asteroid Belt: Rocky fragments from the early solar system between Mars and Jupiter.
- Comets: Icy visitors from the outer regions, dazzling with bright comas and tails.
- Dwarf Planets: Pluto and others, mainly in the Kuiper Belt, challenging our definitions of planets.

Journey Through the Cosmos

Explore the structure of our solar system, the forces at play, and the fascinating variety of bodies that orbit our Sun. This journey invites you to learn, color, and appreciate the vastness and beauty of our place in the cosmos.

Chapter2 : The Stars

Beacons of the Night Sky
Beacons of the Night Sky

Stars, the beacons of the night sky, are massive celestial bodies made primarily of hydrogen and helium. Through the process of nuclear fusion, stars shine brightly, emitting light and heat across vast distances. Stars vary widely in size, color, and lifespan, ranging from red dwarfs that burn slowly but persist for billions of years, to massive blue giants that live fast and die young.

Formation and Lifecycle of Stars
Formation and Lifecycle of Stars

Stars form in vast clouds of gas and dust known as nebulae. When regions within these nebulae collapse under gravity, they form protostars. As the protostar's core temperature increases, nuclear fusion ignites, transforming it into a main-sequence star. This phase can last millions to billions of years, depending on the star's mass.

- Nuclear Fusion: The process by which stars produce energy. In the core of a star, hydrogen atoms fuse to form helium, releasing immense amounts of energy in the form of light and heat. This energy radiates outward, balancing the gravitational forces pulling the star inward.
- Main Sequence: This is the longest phase in a star's life, where it fuses hydrogen into helium in its core. Our Sun is currently in the main sequence phase.
- Post-Main Sequence: Once the hydrogen in the core is exhausted, stars enter the next phase of their evolution. This phase differs significantly between low-mass and high-mass stars.

Low-Mass Stars: These stars, including red dwarfs, live for billions of years due to their efficient use of hydrogen fuel. When they exhaust their hydrogen, they become red giants, expanding and cooling their outer layers. Eventually, they shed these layers, leaving behind a hot core that becomes a white dwarf. Over time, white dwarfs cool and fade away as black dwarfs, a theoretical final state that has not yet been observed.

High-Mass Stars: Stars with much greater mass undergo more dramatic transformations. After exhausting their hydrogen, they become red supergiants and begin fusing heavier elements in their cores. This process continues until iron is formed, which cannot produce energy through fusion.

The core collapses, leading to a supernova explosion, one of the most powerful events in the universe. The remnants can form neutron stars or black holes.

Explore the Colorful Diversity of Stars

Stars come in various colors, indicating their temperature and stage in the stellar lifecycle. This activity will help you understand and visualize the diversity among stars:

·Red Giants: These are aging stars that have expanded and cooled, appearing red. Red giants are much larger and cooler than their main-sequence counterparts. Famous examples include Betelgeuse in the constellation Orion and Antares in Scorpius.

·White Dwarfs: The remnants of stars like our Sun after they have exhausted their nuclear fuel. White dwarfs are incredibly dense; a teaspoon of white dwarf material would weigh tons on Earth. They are hot but very small and faint compared to main-sequence stars.

·Supernovae: The spectacular explosions that mark the death of massive stars. A supernova can outshine an entire galaxy for a short period and leave behind exotic remnants such as neutron stars or black holes. Famous supernovae include SN 1987A in the Large Magellanic Cloud.

Life Cycle of a Star

Stars are not eternal; they are born, live out their lives, and eventually die, leaving behind fascinating remnants. The journey begins in a nebula, where clouds of gas and dust collapse under gravity to form a new star. As the star ages, it passes through phases like the main sequence, red giant or supergiant, and ultimately may become a supernova.

Educational Focus: Nuclear fusion, the power source of stars, involves hydrogen atoms fusing to form helium under immense pressure and temperature. This process releases a tremendous amount of energy, which we see as light and feel as heat from our Sun.

Exploring Galaxies

Our universe is filled with galaxies, each a grand collection of billions of stars bound together by gravity. Galaxies are not only vast assemblages of stars but also include gas, dust, and dark matter. They come in various shapes and sizes, and their forms and structures provide valuable insights into the processes of galaxy formation and evolution.

Spiral Galaxies

Characteristics:
- Structure: Spiral galaxies, such as our Milky Way, have a flat, disk-like shape with a central bulge surrounded by rotating spiral arms. These arms are sites of active star formation, where new stars are born from dense clouds of gas and dust.
- Components: The central bulge contains older, redder stars and a supermassive black hole, while the spiral arms contain younger, bluer stars, nebulae, and star clusters.
- Examples: The Milky Way, Andromeda (M31), and the Whirlpool Galaxy (M51) are notable spiral galaxies.
- Star Formation: Spiral galaxies are known for their ongoing star formation, primarily occurring in the spiral arms. These regions are rich in interstellar material that collapses under gravity to form new stars.

Subtypes:
- Barred Spiral Galaxies: A subtype of spiral galaxies that feature a central bar structure extending from the nucleus, with spiral arms winding out from the ends of the bar. The Milky Way is an example of a barred spiral galaxy.

Elliptical Galaxies

Characteristics:
- Structure: Elliptical galaxies are massive, rounded clusters of stars with an elliptical shape. They lack the distinct spiral arms found in spiral galaxies and have a more uniform, featureless appearance.
- Components: These galaxies contain older, low-mass stars and very little interstellar gas and dust, indicating a lack of recent star formation.
- Examples: M87 in the Virgo Cluster and NGC 1132 are prominent elliptical galaxies.
- Star Formation: Elliptical galaxies have little to no star formation activity. They are composed of older, evolved stars that formed during the galaxy's early history.

Subtypes:
- Giant Ellipticals: Among the largest galaxies in the universe, giant elliptical galaxies are often found at the centers of galaxy clusters and can contain trillions of stars.
- Dwarf Ellipticals: Smaller and less massive than giant ellipticals, dwarf elliptical galaxies are more common and often found as satellites of larger galaxies.

Irregular Galaxies

Characteristics:
- Structure: Irregular galaxies lack a distinct, regular shape and often appear chaotic, with no clear structure or symmetry. They are rich in gas and dust, which fuels star formation.

- **Components:** These galaxies contain a mix of old and young stars, star clusters, and nebulae. Their irregular shapes are often the result of gravitational interactions or collisions with other galaxies.
- **Examples:** The Large Magellanic Cloud (LMC) and the Small Magellanic Cloud (SMC) are well-known irregular galaxies.
- **Star Formation:** Irregular galaxies are sites of significant star formation activity, with new stars forming in dense regions of gas and dust.

Subtypes:

- **Dwarf Irregulars:** Smaller versions of irregular galaxies, dwarf irregulars are often found as satellite galaxies of larger galaxies and exhibit active star formation despite their small size.

Additional Information on Galaxies

Galaxy Clusters:

- **Definition:** Galaxies are not isolated but often found in groups or clusters. Galaxy clusters contain hundreds to thousands of galaxies bound together by gravity, forming some of the largest structures in the universe.
- **Examples:** The Virgo Cluster, the Coma Cluster, and the Fornax Cluster are prominent examples of galaxy clusters. These clusters are dominated by massive elliptical galaxies and hot intracluster gas.

Interacting Galaxies:

- **Interactions:** Galaxies can interact and merge, leading to dramatic changes in their structure and star formation activity. These interactions can trigger bursts of star formation and lead to the formation of new galaxy types.
- **Examples:** The Antennae Galaxies and the Whirlpool Galaxy with its companion NGC 5195 are examples of interacting galaxies.

Galactic Halos and Dark Matter:

- **Halos:** Galaxies are surrounded by halos of dark matter, which do not emit light but exert gravitational forces. These halos extend far beyond the visible parts of galaxies and play a crucial role in their dynamics and evolution.
- **Dark Matter:** Dark matter is an essential component of galaxies, making up most of their mass. Its presence is inferred from the gravitational effects on visible matter and the rotation curves of galaxies.

Galaxy Formation and Evolution:

- **Formation:** Galaxies formed from primordial gas clouds that collapsed under gravity in the early universe. Over time, they evolved through processes such as star formation, supernova explosions, and interactions with other galaxies.
- **Evolution:** The study of distant galaxies reveals how they have evolved over billions of years. Observations from telescopes like Hubble provide snapshots of galaxies at different stages of their evolution, helping us understand the history of the universe.

Fun Facts About Galaxies

- **Size and Scale:** The Milky Way, our home galaxy, is about 100,000 light-years in diameter and contains over 200 billion stars. Andromeda, the nearest spiral galaxy to the Milky Way, is about 2.5 million light-years away and on a collision course with the Milky Way, expected to merge in about 4.5 billion years.
- **Quasars:** Some galaxies have extremely bright centers known as quasars, powered by supermassive black holes that actively accrete matter. Quasars are among the brightest objects in the universe.
- **Galactic Cannibalism:** Larger galaxies can grow by absorbing smaller galaxies in a process known as galactic cannibalism. This process contributes to the growth and evolution of galaxies over time.

By exploring the diverse types of galaxies and their fascinating characteristics, you gain a deeper appreciation for the complexity and grandeur of the universe. Each galaxy, with its unique history and properties, contributes to the rich tapestry of the cosmos.

Deep Space Objects

Beyond galaxies, the universe is teeming with exotic objects:
- **Quasars:** The ultra-bright centers of distant galaxies, powered by supermassive black holes.
- **Pulsars:** Neutron stars that emit beams of radiation like cosmic lighthouses.
- **Black Holes:** Regions of space where gravity is so strong, not even light can escape.

Chapter 3: Astronomical Phenomena

Meteors, Meteorites, and Comets: Cosmic Rockstars!

Meteors: The Sky's Shooting Stars

Shooting Stars: Ever wished upon a shooting star? Meteors, often called shooting stars, are bits of rock or dust that blaze across the sky, creating a brief but brilliant streak of light. These celestial fireworks are actually meteoroids burning up as they enter Earth's atmosphere.

Formation: Meteors are the visible paths of meteoroids—tiny particles from comets or asteroids. When these particles zoom into Earth's atmosphere at high speeds, friction with the air makes them heat up and glow. Voilà, a shooting star!

Types of Meteors:

- Sporadic Meteors: These are the random, unexpected shooting stars you might see on any clear night.
- Meteor Showers: Predictable annual events where multiple meteors radiate from a specific point in the sky, making it look like the universe is having a party.

Meteorites: Space Rocks on Earth

Surviving the Journey: When a meteor survives its fiery fall and lands on Earth, it's called a meteorite. These space rocks provide precious clues about the solar system's history and composition.

Types of Meteorites:
- Stony Meteorites: The most common type, composed mainly of silicate minerals.
- Iron Meteorites: Dense, metallic rocks made primarily of iron and nickel.
- Stony-Iron Meteorites: A rare blend of metal and silicate minerals.

Scientific Importance: Meteorites are like time capsules from space, revealing secrets about the early solar system and the building blocks of planets. Some even contain organic compounds, sparking curiosity about the origins of life.

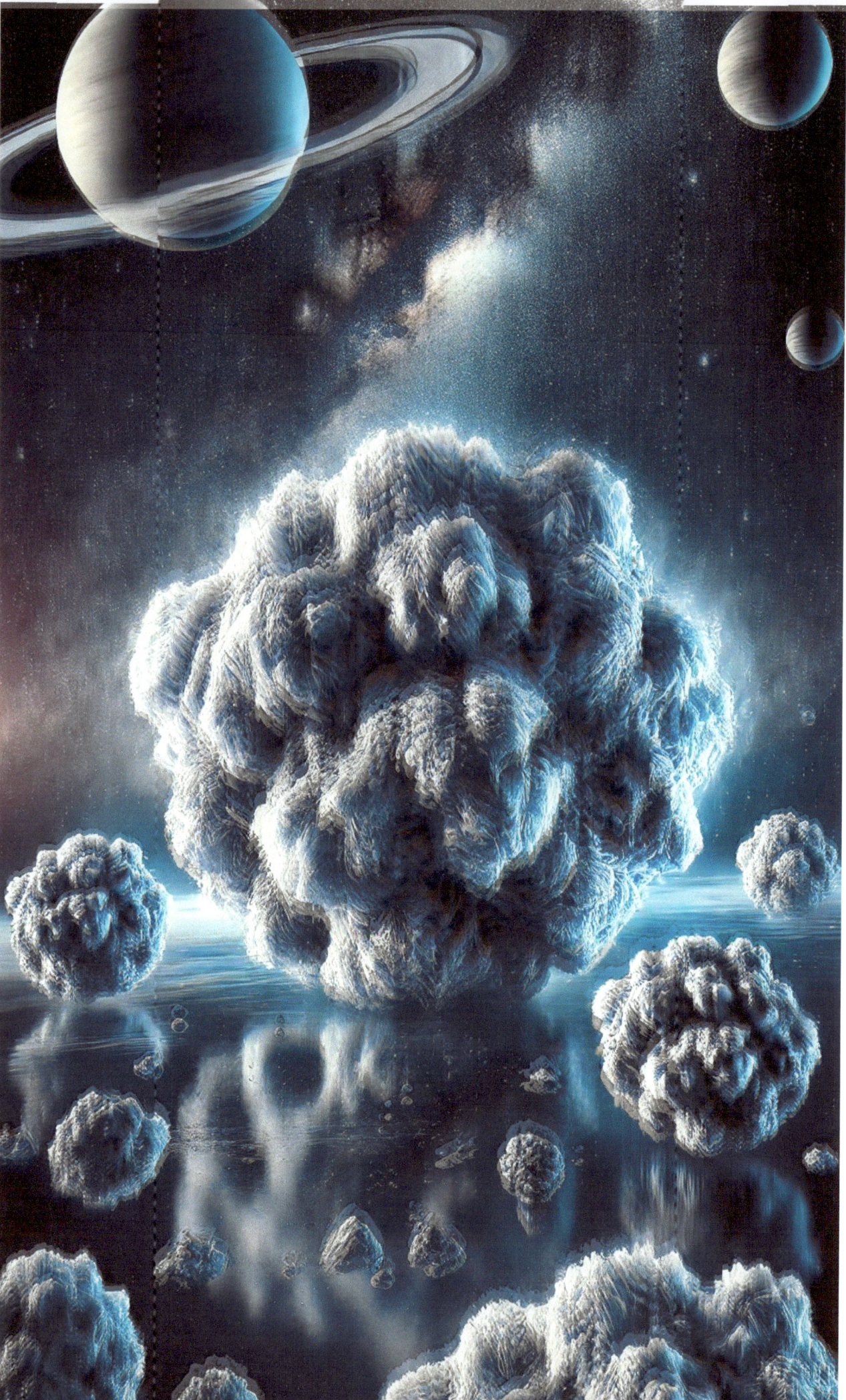

Icy Bodies: Cosmic Icebergs
Icy Bodies: Cosmic Icebergs

Comets are like cosmic icebergs from the far-off Kuiper Belt and Oort Cloud. When they near the Sun, their icy surfaces vaporize, forming glowing comas and spectacular tails that stretch millions of kilometers.

Comet Structure:
- **Nucleus:** The solid core made of rock, dust, water ice, and frozen gases.
- **Coma:** A fuzzy, glowing cloud of gas and dust that appears as the comet heats up.
- **Tails:** Comets have two tails—a bright ion tail made of ionized gas and a dusty tail. These tails always point away from the Sun because of the solar wind and radiation pressure.

Cometary Orbits:
Comets follow long, stretched-out orbits that take them close to the Sun and then fling them far back into the outer solar system. Some make this journey in a few years, while others take thousands.

Where Do They Come From?
Origins of Meteors and Meteorites:
- **Debris from Comets and Asteroids:** Most meteoroids come from comet debris or the asteroid belt between Mars and Jupiter.
- **Meteor Showers:**
  - **Perseids:** Light up the sky every August from Comet Swift-Tuttle's debris.
  - **Leonids:** Occur every November, thanks to Comet Tempel-Tuttle.
  - **Geminids:** Dazzle us every December, linked to asteroid 3200 Phaethon.

Origins of Comets:

- **Kuiper Belt:** A disk-shaped region beyond Neptune filled with short-period comets.
- **Oort Cloud:** A distant, spherical shell around the solar system, home to long-period comets with orbits spanning thousands to millions of years.

Famous Examples
Famous Examples

Halley's Comet
- **Visibility:** Halley's Comet makes an appearance every 76 years, a true cosmic celebrity.
- **Historical Significance:** This comet has been recorded for centuries, linked to events like the Battle of Hastings in 1066.
- **Scientific Observations:** The Giotto spacecraft provided close-up views in 1986.

Chelyabinsk Meteor:
- Event: In 2013, a large meteor exploded over Chelyabinsk, Russia, with the energy of a small nuclear bomb.
- Impact: The blast shattered windows and injured around 1,500 people.
- Scientific Insights: Highlighted the potential hazards of near-Earth objects.

Additional Information

Meteor Showers:
- Annual Events: Meteor showers are predictable and provide a dazzling display as Earth passes through comet debris.
- Observation Tips: For the best view, find a dark, clear sky away from city lights, usually after midnight when the radiant point is higher in the sky.

Impact Craters:
- Formation: Larger meteoroids or asteroids striking Earth create impact craters.
- Famous Craters: The Barringer Crater in Arizona and the Chicxulub Crater in Mexico. The latter is linked to the dinosaur extinction event.

Cometary Missions:
- Rosetta Mission: Orbited and landed on Comet 67P/Churyumov-Gerasimenko, giving us an up-close look at a comet's surface.
- Deep Impact Mission: NASA's Deep Impact mission sent a probe to collide with Comet Tempel 1, providing insights into its composition.

Eclipses and Auroras: Cosmic Light Shows

Solar Eclipses: The Ultimate Cosmic Hide-and-Seek

Solar Eclipse: Imagine the Moon playing a giant game of hide-and-seek with the Sun! A solar eclipse happens when the Moon passes between Earth and the Sun, blocking some or all of the Sun's light. This alignment causes the Moon's shadow to fall on Earth, leading to a temporary darkening of the sky.

Types of Solar Eclipses:
- Total Solar Eclipse:
  - What Happens: The Moon completely covers the Sun.
  - Cool Factor: The sky darkens like it's twilight, and you see the Sun's corona, a glowing halo of plasma.
  - Duration: Totality can last up to 7.5 minutes but is typically shorter.
  - Rarity: These events are rare for any given location on Earth. In fact, the same spot on Earth only gets to see a total solar eclipse once every 360 to 410 years!

- Partial Solar Eclipse:
  - What Happens: Only a portion of the Sun is covered by the Moon.
  - Cool Factor: The Sun looks like it has a bite taken out of it, creating a crescent shape.
  - Commonness: More common than total eclipses and visible over larger areas.
- Annular Solar Eclipse:
  - What Happens: The Moon is too far from Earth to completely cover the Sun, leading to a "ring of fire" effect.
  - Cool Factor: You get a stunning ring-like appearance where the outer edge of the Sun is still visible around the Moon.
  - Visuals: This type of eclipse happens when the Moon is near apogee, the farthest point in its orbit from Earth.
-

Observation and Safety:
- Safety First: Never look directly at a solar eclipse without special solar viewing glasses or a pinhole projector. Regular sunglasses won't protect your eyes and can cause serious eye damage.
- Viewing Methods: Use eclipse glasses, welder's glasses with a shade of 14, or make a pinhole projector to watch the eclipse safely.

Lunar Eclipses: The Moon's Shadow Dance

Lunar Eclipse: A lunar eclipse happens when Earth passes between the Sun and the Moon, casting a shadow on the Moon. This can only occur during a full moon when the Sun, Earth, and Moon are in near-perfect alignment.

Types of Lunar Eclipses:
- Total Lunar Eclipse:
  - What Happens: The entire Moon passes through Earth's umbra, often turning a reddish color due to Rayleigh scattering of sunlight through Earth's atmosphere—a phenomenon known as a "blood moon."
  - Cool Factor: The Moon looks eerily beautiful in shades of red, orange, or brown.
  - Duration: Totality can last from a few minutes to over an hour.
- Partial Lunar Eclipse:
  - What Happens: Only a portion of the Moon enters Earth's umbra, creating a noticeable dark shadow on part of the Moon's surface.
  - Cool Factor: It looks like someone took a cosmic bite out of the Moon!
- Penumbral Lunar Eclipse:
  - What Happens: The Moon passes through Earth's penumbra, causing a subtle shading that is often difficult to see without precise instruments.
  - Cool Factor: The effect is very subtle, almost like a slight smudge on the Moon's face.

- Observation:
- Eye-Friendly: Unlike solar eclipses, lunar eclipses are safe to view with the naked eye. You can enjoy them without any special equipment.
- Wide Coverage: They can be seen from anywhere on Earth where the Moon is above the horizon during the eclipse, making them more accessible to viewers worldwide.

Auroras: The Sky's Neon Lights

Auroras: Also known as the Northern Lights (Aurora Borealis) and Southern Lights (Aurora Australis), these natural light displays are like Earth's own neon signs, seen predominantly in high-latitude regions around the Arctic and Antarctic. They occur when charged particles from the Sun collide with gases in Earth's atmosphere, causing the gases to emit light.

Formation:
- Solar Wind: The Sun emits a stream of charged particles known as the solar wind. When these particles reach Earth, they interact with its magnetic field.
- Magnetic Field: Earth's magnetic field guides the charged particles toward the poles. As the particles collide with atoms and molecules in the atmosphere, primarily oxygen and nitrogen, they transfer energy, causing these atoms to emit light.

Colors and Variations:
- Common Colors: The most common auroral colors are green and pink. Green is produced by oxygen molecules located about 60 miles above the Earth. Pink (and a mix of green and red) can be seen when both high-altitude oxygen and nitrogen molecules are excited.
- Rare Colors: Red, yellow, blue, and violet auroras can also appear. Red auroras are produced by high-altitude oxygen, at heights of up to 200 miles. Blue and purple hues are produced by nitrogen.
- Patterns: Auroras can take various forms, including arcs, curtains, rays, and coronas. These shapes can shift and change rapidly, creating a mesmerizing display.

Best Viewing Conditions:
- Locations: Auroras are best viewed in high-latitude regions, such as Scandinavia, Canada, Alaska, and Antarctica. They are typically visible during winter months when the nights are longer and skies are clearer.
- Timing: The best times to observe auroras are during periods of high solar activity, known as solar maximum, which occurs roughly every 11 years. Clear, dark nights away from city lights provide the best viewing conditions.

- Observation:
- Eye-Friendly: Unlike solar eclipses, lunar eclipses are safe to view with the naked eye. You can enjoy them without any special equipment.
- Wide Coverage: They can be seen from anywhere on Earth where the Moon is above the horizon during the eclipse, making them more accessible to viewers worldwide.

Auroras: The Sky's Neon Lights

Auroras: Also known as the Northern Lights (Aurora Borealis) and Southern Lights (Aurora Australis), these natural light displays are like Earth's own neon signs, seen predominantly in high-latitude regions around the Arctic and Antarctic. They occur when charged particles from the Sun collide with gases in Earth's atmosphere, causing the gases to emit light.

Formation:
- Solar Wind: The Sun emits a stream of charged particles known as the solar wind. When these particles reach Earth, they interact with its magnetic field.
- Magnetic Field: Earth's magnetic field guides the charged particles toward the poles. As the particles collide with atoms and molecules in the atmosphere, primarily oxygen and nitrogen, they transfer energy, causing these atoms to emit light.

Colors and Variations:
- Common Colors: The most common auroral colors are green and pink. Green is produced by oxygen molecules located about 60 miles above the Earth. Pink (and a mix of green and red) can be seen when both high-altitude oxygen and nitrogen molecules are excited.
- Rare Colors: Red, yellow, blue, and violet auroras can also appear. Red auroras are produced by high-altitude oxygen, at heights of up to 200 miles. Blue and purple hues are produced by nitrogen.
- Patterns: Auroras can take various forms, including arcs, curtains, rays, and coronas. These shapes can shift and change rapidly, creating a mesmerizing display.

Best Viewing Conditions:
- Locations: Auroras are best viewed in high-latitude regions, such as Scandinavia, Canada, Alaska, and Antarctica. They are typically visible during winter months when the nights are longer and skies are clearer.
- Timing: The best times to observe auroras are during periods of high solar activity, known as solar maximum, which occurs roughly every 11 years. Clear, dark nights away from city lights provide the best viewing conditions.

- Geomagnetic Storms: Intense auroras are often associated with geomagnetic storms, which occur when the solar wind is particularly strong. These storms can cause auroras to be visible at lower latitudes than usual.

Scientific and Cultural Significance

Studying Auroras:
- Scientific Value: Auroras provide scientists with valuable information about Earth's magnetic field and the behavior of the solar wind. Observing auroras helps researchers understand space weather and its potential impacts on satellite communications and power grids.
- Cultural Importance: Throughout history, auroras have inspired myths and legends in various cultures. Indigenous peoples of the Arctic, for example, have rich folklore surrounding the Northern Lights, viewing them as spiritual or mystical phenomena.

By exploring the fascinating phenomena of meteors, meteorites, comets, eclipses, and auroras, we gain a deeper appreciation for the dynamic interactions between celestial bodies and Earth's atmosphere. These events not only offer spectacular visual displays but also enhance our understanding of the natural world and the forces that shape it.

Discovering New Worlds

Exoplanets and Alien Worlds
Exoplanets, or extrasolar planets, are planets that orbit stars outside our solar system. Discovering these distant worlds has revolutionized our understanding of planetary systems and the potential for life beyond Earth. With advanced telescopes and space missions, astronomers have discovered thousands of exoplanets, each with its own unique characteristics.

Methods of Discovery:
Transit Method:
- How It Works: Imagine a tiny planet crossing in front of a big, bright star, causing the star to dim a little. That's what happens in the transit method! This dimming tells astronomers that a planet is there.
- Kepler Space Telescope: This telescope was like a super spy, watching over 150,000 stars and finding more than 2,600 planets!
- Advantages: It helps us figure out how big the planet is and how long it takes to orbit its star.
- Limitations: The planet, star, and observer need to line up perfectly. If not, we might miss seeing the planet.

Radial Velocity Method:
- How It Works: Planets make their stars wobble a little as they orbit. By watching this wobble, we can find planets! It's like watching a dance between the star and its planet.
- Spectroscopic Analysis: By examining the star's light, we can see the tiny shifts caused by the planet's pull.
- Pioneering Discoveries: This method helped discover the first planet around a star like our Sun in 1995.
- Advantages: It's great for finding planets that don't transit their stars.
- Limitations: It works best for big planets close to their stars. Small, Earth-like planets are harder to find this way.

Direct Imaging:
- How It Works: Imagine trying to see a firefly next to a lighthouse. Direct imaging involves blocking the star's light to see the planet next to it.
- Challenges: Planets are much dimmer than their stars and often very close to them, making it tricky.
- Successful Missions: The Very Large Telescope (VLT) and the Gemini Planet Imager have snapped some cool planet pics!
- Advantages: We can study the planet's atmosphere and potential for life.
- Limitations: Right now, it's mostly good for spotting big planets far from their stars, but we're getting better!

Famous Exoplanets

Kepler-186f:
- Description: This rocky planet is about the same size as Earth and orbits in the habitable zone of its star, Kepler-186, which means it might have liquid water.
- Significance: Discovered in 2014, it was the first Earth-sized planet found in the habitable zone of another star.
- Potential for Life: Liquid water is key for life as we know it, making Kepler-186f a fascinating place to study.

TRAPPIST-1e:
- Description: One of seven Earth-sized planets orbiting the cool star TRAPPIST-1, located about 39 light-years away.
- Significance: Discovered in 2017, this system is famous for having many Earth-sized planets.
- Potential for Life: TRAPPIST-1e is especially interesting because it's in the habitable zone, where conditions might support life.

Additional Methods of Exoplanet Discovery

Gravitational Microlensing:
- How It Works: This method uses the gravitational pull of a star and its planet to magnify the light of a more distant star, revealing the planet.
- Advantages: It can find planets far from their stars, even in remote parts of the galaxy.
- Limitations: These events are rare and unpredictable, so follow-up observations are tough.

Astrometry:
- How It Works: This involves measuring the precise movements of a star caused by a planet's gravitational pull.
- Advantages: It can detect planets in wide orbits and provide direct measurements of a planet's path.
- Limitations: Requires extremely high precision and can be affected by other motions of the star.

Discovering exoplanets is like a cosmic treasure hunt, revealing the incredible diversity of worlds beyond our solar system. From rocky Earth-like planets to gas giants, each discovery brings us closer to understanding our place in the universe and the potential for life beyond our own planet.

The Search for Habitable Worlds
The Search for Habitable Worlds

Habitability Criteria:
- Habitable Zone: The region around a star where conditions might be right for liquid water to exist on a planet's surface. This zone varies depending on the star's size and temperature.
- Atmospheric Conditions: The presence and composition of an atmosphere are crucial for maintaining surface temperatures and protecting potential life from harmful radiation.
- Planetary Composition: Rocky planets are considered more likely to be habitable than gas giants. The presence of water, organic molecules, and stable climates are key factors.

Future Missions and Technologies:
- James Webb Space Telescope (JWST): Scheduled to launch soon, JWST will have the capability to study the atmospheres of exoplanets, searching for signs of habitability and possibly even life.
- Extremely Large Telescopes (ELTs): Ground-based telescopes like the Extremely Large Telescope (ELT) and the Giant Magellan Telescope (GMT) will provide unprecedented resolution and sensitivity for direct imaging and characterization of exoplanets.
- Space Interferometry: Future missions may use interferometry, combining the light from multiple telescopes to achieve higher resolution images, allowing for the detection of smaller, Earth-like exoplanets.

By exploring the methods used to discover exoplanets and studying famous examples, we gain a deeper understanding of the diversity and potential habitability of planetary systems beyond our own. The quest to find Earth-like worlds continues to drive advancements in technology and deepen our knowledge of the universe.

Space Weather

Understanding Space Weather:
Space weather refers to the environmental conditions in space as influenced by the Sun and the solar wind. It encompasses various phenomena such as solar flares, geomagnetic storms, and cosmic rays. Understanding space weather is crucial for protecting both our technological infrastructure on Earth and human activities in space.

Key Phenomena
Solar Winds:

- Definition: Solar winds are streams of charged particles, primarily electrons and protons, released from the upper atmosphere of the Sun, known as the corona. These particles travel through space at speeds ranging from 250 to 750 kilometers per second (155 to 466 miles per second).
- Formation: The solar wind is generated by the high temperatures in the Sun's corona, which give particles enough energy to escape the Sun's gravitational pull.
- Impact on Earth: When the solar wind reaches Earth, it interacts with our planet's magnetic field, causing geomagnetic storms and auroras. Solar winds can also compress Earth's magnetosphere, affecting satellites and communications systems.

Cosmic Rays:

- Definition: Cosmic rays are high-energy particles originating from outside the solar system. They consist primarily of protons, but also include heavier atomic nuclei and electrons.
- Sources: Cosmic rays are produced by supernova explosions, active galactic nuclei, and other high-energy astrophysical processes. They travel at nearly the speed of light and can penetrate the Earth's atmosphere.
- Impact on Space Missions: Cosmic rays pose a significant risk to astronauts by increasing their exposure to radiation, which can lead to health issues. They also affect electronic equipment in space, potentially causing malfunctions in satellites and spacecraft.

Geomagnetic Storms:
- Definition: Geomagnetic storms are disturbances in Earth's magnetosphere caused by interactions with the solar wind, particularly during coronal mass ejections (CMEs), where large amounts of solar plasma and magnetic fields are ejected into space.
- Effects: These storms can cause fluctuations in Earth's magnetic field, leading to disruptions in power grids, communication systems, and navigation systems. They can also produce vivid auroras in high-latitude regions.
- Measurement: Geomagnetic storms are measured using the K-index and A-index, which quantify the disturbance in Earth's magnetic field. Severe geomagnetic storms can have significant economic and technological impacts.

Solar Flares:
- Definition: Solar flares are sudden, intense bursts of energy and radiation from the Sun's surface, associated with sunspots and magnetic activity. They release energy equivalent to millions of hydrogen bombs.
- Classification: Solar flares are classified based on their X-ray brightness, with classes ranging from A, B, C, M, to X, with X being the most powerful.
- Impact on Earth: Solar flares can disrupt satellite communications, GPS signals, and power grids. They can also pose a radiation hazard to astronauts and high-altitude flights, particularly over polar regions.

Effects on Earth

Geomagnetic Storms:
- Auroras: One of the most visible effects of geomagnetic storms is the production of auroras, also known as the Northern and Southern Lights. These natural light displays occur when charged particles from the solar wind interact with Earth's magnetic field and atmosphere.
- Power Grids: Geomagnetic storms can induce electric currents in power lines, potentially causing voltage instability and transformer damage. This can lead to widespread power outages and equipment failures.
- Navigation Systems: Disruptions in the Earth's magnetic field can affect navigation systems, including GPS and aircraft navigation. This can lead to inaccuracies and increased risks for transportation and military operations.

Solar Flares:
- Satellite Communications: The radiation from solar flares can affect the ionosphere, the layer of Earth's atmosphere that reflects and modifies radio waves used for communication. This can result in signal degradation or complete loss of communication with satellites.
- GPS Signals: GPS satellites rely on accurate timing signals that can be disrupted by solar flares, leading to positioning errors. This impacts navigation for aircraft, ships, and even personal GPS devices.

- **Radiation Hazards**: Increased radiation levels from solar flares can pose a threat to astronauts, particularly those on the International Space Station (ISS) or on future missions to the Moon and Mars. High-altitude flights over polar regions may also experience increased radiation exposure.

Monitoring and Predicting Space Weather:

- **Space Weather Centers**: Organizations like NOAA's Space Weather Prediction Center (SWPC) and NASA's Heliophysics Division monitor solar activity and provide forecasts and warnings about space weather events.
- **Satellites**: Satellites such as the Solar and Heliospheric Observatory (SOHO), the Solar Dynamics Observatory (SDO), and the Advanced Composition Explorer (ACE) continuously observe the Sun and its interactions with the Earth's environment.
- **Predictive Models**: Scientists use computer models to predict the impact of solar activity on Earth's space weather. These models help forecast the arrival and potential effects of solar storms, aiding in the preparation and mitigation of their impacts.

Mitigation Strategies:

- **Satellite Design**: Satellites are built with shielding and redundant systems to protect against space weather effects. Engineers design satellites to withstand radiation and continue functioning during geomagnetic storms.
- **Power Grid Protections**: Power companies implement strategies to protect the grid, such as installing transformers with geomagnetic shielding, developing rapid response protocols, and creating backup systems to manage the load during geomagnetic storms.
- **Astronaut Safety**: Space agencies monitor space weather to protect astronauts by providing warnings and instructions to seek shelter in more shielded parts of the spacecraft during solar radiation storms.

Historical Space Weather Events:

- **Carrington Event (1859)**: The most powerful geomagnetic storm on record, caused by a massive solar flare and CME. It led to widespread telegraph disruptions and auroras visible as far south as the Caribbean.
- **March 1989 Geomagnetic Storm**: A powerful CME caused a geomagnetic storm that knocked out power to millions of people in Quebec, Canada, and damaged satellites.
- **Halloween Storms (2003)**: A series of intense solar storms in late October and early November 2003 disrupted satellite communications, caused power outages in Sweden, and created vivid auroras visible at lower latitudes than usual.

Understanding space weather is essential for maintaining the safety and functionality of our technological systems and human activities in space. As our reliance on technology grows and human presence in space expands, the need for advanced space weather monitoring and prediction becomes increasingly important.

Chapter 4: Modern Astronomy

Telescopes - The Astronomer's Eye & Windows to the Universe

Telescopes have revolutionized our understanding of the universe, allowing us to peer into the depths of space and observe distant stars, planets, and galaxies. From the early refracting telescopes used by Galileo to the advanced space telescopes orbiting Earth today, these instruments have been pivotal in our quest to understand the cosmos. The development and refinement of telescopes have continually expanded our view, revealing the vastness and complexity of the universe.

Types of Telescopes

Refracting Telescopes:
- How They Work: Refracting telescopes use lenses to bend (refract) and focus light. The main lens, called the objective lens, gathers light and brings it to a focus, creating an image that is viewed through an eyepiece.
- Historical Significance: Galileo Galilei was the first to use a refracting telescope for astronomical observations in the early 1600s. With his telescope, Galileo discovered the four largest moons of Jupiter, observed the phases of Venus, and studied sunspots and the rugged surface of the Moon.
- Limitations: Refracting telescopes suffer from chromatic aberration, where different colors of light are focused at different points, causing a halo effect around objects. Large lenses are also difficult and expensive to produce and support.
- Modern Use: While large refracting telescopes are no longer commonly used in professional astronomy, they are still popular among amateur astronomers for their simplicity and clear, sharp images.

Reflecting Telescopes:
- How They Work: Reflecting telescopes use mirrors to gather and focus light. The primary mirror collects light and reflects it to a focus point, where a secondary mirror may redirect it to an eyepiece or a detector.
- Invention and Development: Isaac Newton invented the reflecting telescope in 1668 to overcome the chromatic aberration problem of refractors. Reflecting telescopes can be built larger than refractors, allowing them to collect more light and observe fainter objects.
- Types of Reflecting Telescopes:
  - Newtonian Reflector: Features a primary concave mirror and a flat diagonal secondary mirror.
  - Cassegrain Reflector: Uses a concave primary mirror and a convex secondary mirror, directing light through a hole in the primary mirror to the eyepiece or detector.

- Significant Reflectors: The Hale Telescope at Palomar Observatory and the Keck Telescopes in Hawaii are among the largest and most powerful reflecting telescopes in the world.

Radio Telescopes:
- How They Work: Radio telescopes detect radio waves emitted by celestial objects. They use large parabolic dishes to collect and focus radio waves onto a receiver.
- Importance in Astronomy: Radio telescopes have opened up a new window to the universe, allowing astronomers to study phenomena that are not visible in optical light, such as pulsars, quasars, and the cosmic microwave background radiation.
- Key Observatories: Notable radio telescopes include the Arecibo Observatory (before its collapse), the Very Large Array (VLA) in New Mexico, and the Parkes Observatory in Australia.
- Interferometry: Multiple radio telescopes can be linked together to form an interferometer, providing higher resolution images. The Event Horizon Telescope, which captured the first image of a black hole, is an example of such a network.

Space Telescopes:
- How They Work: Space telescopes are positioned above Earth's atmosphere to avoid the distortion and absorption of light caused by air turbulence and atmospheric particles. This allows them to capture clearer and more detailed images.
- Hubble Space Telescope: Launched in 1990, Hubble has provided some of the most stunning and detailed images of the universe. It has helped determine the rate of expansion of the universe, observed distant galaxies, and studied the atmospheres of exoplanets.

- Other Notable Space Telescopes:
  - James Webb Space Telescope (JWST): Set to succeed Hubble, JWST will observe in the infrared spectrum, allowing it to study the early universe, the formation of stars and planets, and the atmospheres of potentially habitable exoplanets.
  - Chandra X-ray Observatory: Specializes in X-ray astronomy, observing high-energy regions of the universe, such as the remnants of supernovae, black holes, and galaxy clusters.
  - Spitzer Space Telescope: Observed the universe in infrared light, providing insights into the cold and dusty regions of space, such as star-forming nebulae and the centers of galaxies.

The Evolution of Telescopes

Early Developments:
- Hans Lippershey: Credited with inventing the first practical telescope in 1608, a simple refracting telescope.

- **Galileo's Contributions:** Improved upon Lippershey's design, making significant astronomical discoveries that challenged the geocentric view of the universe.

Advancements in Mirror Technology:

- **Reflecting Telescopes:** The development of reflecting telescopes allowed for larger apertures and greater light-gathering power. Innovations in mirror-making, such as the use of glass coated with reflective materials, have continually improved their performance.
- **Adaptive Optics:** Modern reflecting telescopes use adaptive optics to compensate for atmospheric distortion in real-time, providing clearer and sharper images of celestial objects.

Radio and Space Astronomy:

- **Pioneering Radio Telescopes:** The construction of large radio telescopes in the mid-20th century revolutionized our understanding of the universe, revealing phenomena such as the cosmic microwave background radiation and radio galaxies.
- **Space-based Observatories:** The launch of space telescopes in the latter half of the 20th century has provided unprecedented views of the universe across the entire electromagnetic spectrum, from gamma rays to infrared.

Impact on Our Understanding of the Universe

Expanding the Cosmic Horizon:

- **Deep Field Observations:** Telescopes like Hubble have conducted deep field observations, capturing images of galaxies billions of light-years away and providing a glimpse into the early universe.
- **Dark Matter and Dark Energy:** Observations from telescopes have contributed to the discovery and study of dark matter and dark energy, which make up the majority of the universe's mass-energy content.

Exoplanet Discoveries:

- **Kepler and Beyond:** The Kepler Space Telescope revolutionized the search for exoplanets, discovering thousands of planets and revealing the diversity of planetary systems. Future telescopes like JWST and the Transiting Exoplanet Survey Satellite (TESS) will continue this quest.

Understanding Cosmic Phenomena:

- **Black Holes and Neutron Stars:** Telescopes have provided critical data on extreme objects like black holes and neutron stars, enhancing our understanding of their formation, behavior, and impact on their surroundings.
- **Star Formation and Evolution:** Observations of star-forming regions and stellar nurseries have shed light on the processes that lead to the birth and evolution of stars and planetary systems.

The Future of Telescopic Observations

Next-Generation Telescopes:

- **Extremely Large Telescopes (ELTs):** Ground-based telescopes such as the Extremely Large Telescope (ELT) and the Giant Magellan Telescope (GMT) will have unprecedented light-gathering power and resolution, enabling detailed studies of distant galaxies, exoplanets, and cosmic phenomena.
- **Space Missions:** Upcoming missions like the James Webb Space Telescope (JWST) and the Wide Field Infrared Survey Telescope (WFIRST) promise to revolutionize our understanding of the universe by providing deeper and more detailed observations.

International Collaboration:

- **Global Networks:** The construction and operation of large telescopes often involve international collaboration, pooling resources and expertise from multiple countries to achieve scientific goals.
- **Data Sharing:** Advances in data sharing and computational analysis allow astronomers worldwide to collaborate on the interpretation of observations, leading to rapid advancements in our understanding of the universe.

Telescopes continue to be our most powerful tools for exploring the universe, revealing the wonders of the cosmos and expanding our knowledge of the fundamental processes that govern it. As technology advances, these instruments will undoubtedly uncover even more secrets of the universe, fueling our curiosity and driving scientific discovery.

Important Missions and Discoveries
Important Missions and Discoveries

Over the years, numerous space missions have significantly expanded our knowledge of the solar system and beyond. These missions have provided detailed images, invaluable data, and groundbreaking discoveries that have shaped our understanding of the universe. Here are some key missions that have made significant contributions to astronomy:

Voyager Missions
Voyager Missions

Launch and Objectives:

- **Launch Dates:** Voyager 2 was launched on August 20, 1977, followed by Voyager 1 on September 5, 1977.
- **Primary Mission:** Their primary mission was to explore the outer planets—Jupiter, Saturn, Uranus, and Neptune—and send back detailed images and data.

Achievements:

- Jupiter and Saturn: Voyager 1 and 2 provided the first detailed images of the Jupiter and Saturn systems, revealing their moons, rings, and magnetic fields. Voyager 1 discovered active volcanoes on Jupiter's moon Io, while Voyager 2 captured the intricacies of Saturn's rings and the atmosphere of its moon Titan.
- Uranus and Neptune: Voyager 2 remains the only spacecraft to have visited Uranus and Neptune, providing unprecedented data on their atmospheres, rings, and moons. It discovered 10 new moons and two new rings around Uranus, and observed Neptune's Great Dark Spot and high-speed winds.
- Interstellar Space: After completing their planetary missions, both Voyagers continued to travel outward. Voyager 1 entered interstellar space in August 2012, followed by Voyager 2 in November 2018, providing the first direct measurements of the interstellar medium.

Legacy:

- Golden Records: Each spacecraft carries a Golden Record with sounds and images representing Earth's diversity, intended as a message for any potential extraterrestrial life. These records include music, greetings in multiple languages, and images depicting life and culture on Earth.

Hubble Space Telescope

Launch and Objectives:

- Launch Date: The Hubble Space Telescope was launched on April 24, 1990, aboard the Space Shuttle Discovery.
- Primary Mission: Hubble's mission is to observe celestial objects in visible, ultraviolet, and near-infrared light, providing high-resolution images and spectra.

Achievements:

- Deep Field Observations: Hubble's Deep Field and Ultra Deep Field images have revealed thousands of galaxies, some as far as 13.2 billion light-years away, offering a glimpse into the early universe.
- Expansion of the Universe: Observations from Hubble have helped refine the value of the Hubble constant, improving our understanding of the rate of expansion of the universe.
- Exoplanet Studies: Hubble has observed the atmospheres of exoplanets, detecting elements like water vapor and providing insights into their potential habitability.
- Cosmic Phenomena: Hubble has captured stunning images of phenomena such as the Eagle Nebula's Pillars of Creation, the collision of Comet Shoemaker-Levy 9 with Jupiter, and the intricate structures of planetary nebulae.

Legacy:
- Longevity and Impact: Originally designed for a 15-year mission, Hubble has been operational for over three decades, significantly advancing our knowledge of astrophysics and cosmology.

Mars Rovers

Spirit and Opportunity:
- Launch Dates: Spirit was launched on June 10, 2003, and Opportunity on July 7, 2003.
- Primary Mission: To explore the surface of Mars, study its geology, and search for signs of past water activity.

Achievements:
- Spirit: Operated for over six years, discovering evidence of past hydrothermal activity and ancient water.
- Opportunity: Operated for nearly 15 years, providing extensive geological data, including evidence of past liquid water and ancient environments that could have supported microbial life.

Curiosity:
- Launch Date: Curiosity was launched on November 26, 2011.
- Primary Mission: To explore Gale Crater and investigate Mars' climate and geology, assess whether it ever had conditions suitable for microbial life, and study the planet's habitability.

Achievements:
- Mount Sharp: Curiosity has been exploring the lower slopes of Mount Sharp, analyzing sedimentary layers that reveal the history of Mars' climate.
- Organic Molecules: Detected complex organic molecules in Martian rocks, suggesting that the building blocks of life might have been present.

Perseverance:
- Launch Date: Perseverance was launched on July 30, 2020.
- Primary Mission: To search for signs of ancient life, collect samples for future return to Earth, and demonstrate new technologies for human exploration.

Achievements:
- Jezer Crater: Exploring the ancient river delta and lakebed deposits, Perseverance is collecting samples to be cached for future missions.
- Ingenuity Helicopter: Perseverance's mission includes the Ingenuity helicopter, which has successfully conducted powered flights on Mars, demonstrating the feasibility of aerial exploration.

The Future of Space Exploration

The future of space exploration is filled with exciting possibilities. Upcoming missions and advancements promise to take us further than ever before, expanding our understanding of the universe, and paving the way for human exploration of other planets. Here are some of the key programs and missions that will shape the future of space exploration:

Artemis Program
Artemis Program

Objectives:

- Return to the Moon: The Artemis program, led by NASA, aims to return humans to the Moon by 2024. This mission will mark the first time humans have set foot on the Moon since the Apollo missions of the 1960s and 70s.
- Sustainable Presence: The goal is not just to visit the Moon but to establish a sustainable human presence. This involves building the Lunar Gateway, a space station that will orbit the Moon and serve as a staging point for lunar landings.
- Lunar Exploration: Artemis missions will explore the lunar South Pole, a region rich in water ice. This water can be used for life support and potentially converted into rocket fuel, making future missions more sustainable.

Technological Advances:

- Space Launch System (SLS): The most powerful rocket ever built, designed to carry astronauts to the Moon and beyond.
- Orion Spacecraft: A next-generation spacecraft capable of carrying astronauts to deep space destinations, equipped with advanced life support and navigation systems.
- Lunar Gateway: A modular space station that will orbit the Moon, providing support for long-term lunar missions and serving as a testbed for technologies needed for future Mars missions.

Impact on Future Missions:

- Mars Exploration: The Artemis program will lay the groundwork for future human missions to Mars by testing new technologies and techniques on the Moon.
- International Collaboration: Artemis involves international partners, including ESA (European Space Agency), JAXA (Japan Aerospace Exploration Agency), and CSA (Canadian Space Agency), promoting global cooperation in space exploration.

Mars Sample Return
Mars Sample Return

Objectives:

- Sample Collection: NASA's Perseverance rover, currently exploring Mars, is collecting samples of Martian soil and rock. These samples will be cached for future retrieval.

- Return to Earth: The Mars Sample Return mission aims to bring these samples back to Earth for detailed analysis. This will involve multiple missions, including a lander to retrieve the samples and a spacecraft to return them to Earth.

Scientific Significance:
- Planetary Science: Analyzing Martian samples on Earth will provide unprecedented insights into the planet's geology, climate, and potential for past life. Scientists will be able to use advanced laboratory techniques that are not possible with current Mars rovers.
- Astrobiology: The search for signs of past microbial life is a key objective. Organic molecules and biosignatures in the samples could provide evidence of ancient life on Mars.

Technological Challenges:
- Sample Retrieval: The mission will involve a Mars Ascent Vehicle (MAV) to launch the samples from the Martian surface into orbit, a first in space exploration.
- Earth Return: The samples will need to be safely transported back to Earth, requiring a robust containment system to prevent contamination.

James Webb Space Telescope (JWST)

Objectives:
- Exploring the Early Universe: Set to launch soon, the JWST will look further back in time than the Hubble Space Telescope, observing the first galaxies that formed after the Big Bang.
- Infrared Observations: JWST will observe the universe in the infrared spectrum, allowing it to peer through dust clouds and study the formation of stars and planets.

Technological Advances:
- Advanced Optics: JWST features a 6.5-meter primary mirror made of 18 hexagonal segments, providing unprecedented resolution and sensitivity.
- Sunshield: A multi-layer sunshield the size of a tennis court will protect the telescope from the Sun's heat, keeping it at the extremely cold temperatures needed for infrared observations.
- Orbit: JWST will orbit at the second Lagrange point (L2), a stable point in space 1.5 million kilometers from Earth, providing a clear and unobstructed view of the cosmos.

Scientific Impact:
- Exoplanet Research: JWST will study the atmospheres of exoplanets, searching for signs of habitability and potential biosignatures.

- Star and Planet Formation: Observations of protostars and protoplanetary disks will provide insights into the processes that lead to the birth of stars and planetary systems.
- Dark Matter and Dark Energy: JWST will contribute to our understanding of the mysterious components of the universe by studying the large-scale structure of galaxies and galaxy clusters.

Other Upcoming Missions and Programs

Europa Clipper:
- Objective: NASA's Europa Clipper mission, set to launch in the 2020s, will conduct detailed reconnaissance of Jupiter's moon Europa. The mission aims to investigate the moon's ice shell and subsurface ocean, which may harbor conditions suitable for life.
- Significance: Understanding Europa's potential habitability will provide insights into the possibility of life elsewhere in the solar system.

Lunar Gateway:
- Objective: The Lunar Gateway will be a space station orbiting the Moon, supporting long-term lunar exploration and serving as a stepping stone for missions to Mars.
- International Collaboration: Developed in partnership with international space agencies, the Gateway will facilitate scientific research and technology demonstrations crucial for deep space exploration.

Dragonfly:
- Objective: NASA's Dragonfly mission, set to launch in 2027, will send a rotorcraft lander to Saturn's moon Titan. Dragonfly will explore Titan's diverse environments, studying its prebiotic chemistry and potential for life.
- Significance: Titan's thick atmosphere and liquid hydrocarbon lakes make it a unique destination, offering a glimpse into the organic chemistry that may resemble early Earth conditions.

International Lunar Research Station (ILRS):
- Objective: A collaborative project between China and Russia, the ILRS aims to establish a research station on the Moon by the 2030s. The station will support long-term scientific studies and exploration activities.
- Significance: The ILRS will enhance international cooperation in space exploration, contributing to our understanding of the Moon and preparing for future Mars missions.

SpaceX Starship:
- Objective: SpaceX's Starship is designed to be a fully reusable spacecraft capable of carrying humans and cargo to the Moon, Mars, and beyond. Its development is crucial for reducing the cost of space travel and enabling large-scale human settlement of other planets.

- **Star and Planet Formation:** Observations of protostars and protoplanetary disks will provide insights into the processes that lead to the birth of stars and planetary systems.
- **Dark Matter and Dark Energy:** JWST will contribute to our understanding of the mysterious components of the universe by studying the large-scale structure of galaxies and galaxy clusters.

Other Upcoming Missions and Programs

Europa Clipper:
- **Objective:** NASA's Europa Clipper mission, set to launch in the 2020s, will conduct detailed reconnaissance of Jupiter's moon Europa. The mission aims to investigate the moon's ice shell and subsurface ocean, which may harbor conditions suitable for life.
- **Significance:** Understanding Europa's potential habitability will provide insights into the possibility of life elsewhere in the solar system.

Lunar Gateway:
- **Objective:** The Lunar Gateway will be a space station orbiting the Moon, supporting long-term lunar exploration and serving as a stepping stone for missions to Mars.
- **International Collaboration:** Developed in partnership with international space agencies, the Gateway will facilitate scientific research and technology demonstrations crucial for deep space exploration.

Dragonfly:
- **Objective:** NASA's Dragonfly mission, set to launch in 2027, will send a rotorcraft lander to Saturn's moon Titan. Dragonfly will explore Titan's diverse environments, studying its prebiotic chemistry and potential for life.
- **Significance:** Titan's thick atmosphere and liquid hydrocarbon lakes make it a unique destination, offering a glimpse into the organic chemistry that may resemble early Earth conditions.

International Lunar Research Station (ILRS):
- **Objective:** A collaborative project between China and Russia, the ILRS aims to establish a research station on the Moon by the 2030s. The station will support long-term scientific studies and exploration activities.
- **Significance:** The ILRS will enhance international cooperation in space exploration, contributing to our understanding of the Moon and preparing for future Mars missions.

SpaceX Starship:
- **Objective:** SpaceX's Starship is designed to be a fully reusable spacecraft capable of carrying humans and cargo to the Moon, Mars, and beyond. Its development is crucial for reducing the cost of space travel and enabling large-scale human settlement of other planets.

Astronomers at Work: Exploring the Universe

- Astronomers use a variety of tools and techniques to study the universe. Their work involves observing celestial events, analyzing data, and developing theories to explain cosmic phenomena. The life of an astronomer is a blend of scientific inquiry, technological expertise, and creative problem-solving. Here's a detailed look at the daily activities and broader scope of their work.

A Day in the Life of an Astronomer

- Imagine the life of an astronomer, peering into the depths of space, unraveling the mysteries of the cosmos. Their work is a delicate dance of observing celestial events, analyzing vast amounts of data, and developing theories that push the boundaries of our understanding. Let's embark on a journey through a day in the life of these stellar explorers.

Observations:
- Using Telescopes: Picture standing under a clear, starry sky, operating telescopes at observatories like Mauna Kea in Hawaii or the Very Large Telescope in Chile. These ground-based giants, along with space-based marvels like the Hubble Space Telescope, allow astronomers to gather data on stars, planets, and galaxies.
- Remote Observations: Many astronomers now conduct observations from the comfort of their labs, using sophisticated computer systems to control telescopes positioned around the world or in space. This remote capability offers the flexibility to capture fleeting cosmic events.
- Scheduled Observing Time: Access to these telescopes is a prized commodity. Astronomers spend months, sometimes years, planning and proposing their observations, detailing the significance of their research to secure precious observing time.

Data Analysis:
- Processing Images: The raw data collected from telescopes is just the beginning. Astronomers use advanced software to enhance images, remove noise, and correct for distortions caused by Earth's atmosphere, revealing the hidden details of celestial objects.
- Spectroscopy: By splitting light from stars and galaxies into its component colors, astronomers can determine their composition, temperature, velocity, and other properties. This technique is crucial for understanding the physical characteristics of celestial bodies.
- Big Data: Modern astronomy involves handling vast amounts of data. Projects like the Sloan Digital Sky Survey (SDSS) and the upcoming Large Synoptic Survey Telescope (LSST) generate terabytes of data, requiring advanced data processing and machine learning techniques to analyze.

Theoretical Work:
- Developing Models: Theoretical astronomers develop models to explain the behavior of celestial objects and the laws governing the universe. This involves using mathematics and physics to simulate conditions in space and predict phenomena that can be tested through observations.
- Computer Simulations: High-performance computing allows astronomers to simulate complex processes, such as galaxy formation, star evolution, and the dynamics of black holes. These simulations help refine theories and guide observational strategies.
- Collaborative Research: Astronomers often collaborate with colleagues from other disciplines, such as physics, chemistry, and planetary science, to develop comprehensive models that integrate different aspects of cosmic phenomena.

Tools and Techniques

Telescopes:
- Optical Telescopes: These include refracting telescopes, which use lenses, and reflecting telescopes, which use mirrors. They are used to observe visible light from celestial objects.
- Radio Telescopes: These detect radio waves from space, allowing astronomers to study objects like pulsars, quasars, and the cosmic microwave background.
- Space Telescopes: Positioned above Earth's atmosphere, these telescopes avoid atmospheric distortion. Notable examples include Hubble, the Chandra X-ray Observatory, and the James Webb Space Telescope.

Instruments:
- Spectrometers: Used to measure the spectrum of light from astronomical objects, providing information about their composition, temperature, and motion.
- CCD Cameras: Charge-coupled devices (CCDs) capture images with high sensitivity and resolution, essential for both optical and infrared astronomy.
- Adaptive Optics: Systems that correct for the blurring effects of Earth's atmosphere in real-time, providing sharper images of celestial objects.

Data Processing and Analysis:
- Software Tools: Astronomers use specialized software for data reduction, image processing, and spectral analysis. Common tools include IRAF, DS9, and Python libraries like Astropy.
- Machine Learning: Increasingly, machine learning algorithms are used to sift through large datasets, identifying patterns and anomalies that might indicate new discoveries.

Broader Scope of an Astronomer's Work
Broader Scope of an Astronomer's Work

Research and Publications:

- **Academic Papers:** Astronomers publish their findings in scientific journals such as the Astrophysical Journal, Astronomy & Astrophysics, and the Monthly Notices of the Royal Astronomical Society. These publications undergo peer review, ensuring the validity and significance of the research.
- **Conferences and Workshops:** Astronomers present their work at international conferences, such as the American Astronomical Society (AAS) meetings, where they exchange ideas and collaborate on new projects.

Education and Outreach:

- **Teaching:** Many astronomers work at universities and colleges, where they teach courses in astronomy and astrophysics. They mentor undergraduate and graduate students, guiding the next generation of scientists.
- **Public Outreach:** Astronomers engage in public outreach to promote science literacy and share the excitement of their discoveries. This includes giving public lectures, writing popular science articles, and participating in media interviews.
- **Citizen Science:** Projects like Galaxy Zoo and SETI@home involve the public in astronomical research, allowing anyone with internet access to contribute to the discovery process.

Instrument Development:

- **Design and Construction:** Astronomers often collaborate with engineers and technicians to design and build new telescopes and instruments. This involves developing innovative technologies to improve observational capabilities.
- **Testing and Calibration:** Ensuring the accuracy and reliability of astronomical instruments requires rigorous testing and calibration. This is an ongoing process that continues throughout the operational life of the instruments.

Interdisciplinary Work:

- **Astrobiology:** Combining astronomy, biology, and planetary science to search for signs of life beyond Earth. This field explores the conditions necessary for life and identifies potentially habitable exoplanets.
- **Cosmology:** Working with physicists to understand the large-scale structure of the universe, dark matter, and dark energy. Cosmologists use observations of the cosmic microwave background and galaxy distributions to test theories of the universe's origin and evolution.

Challenges and Future Directions

Challenges:

- **Light Pollution:** Urban light pollution hinders ground-based observations. Efforts to establish dark-sky preserves and develop adaptive optics are crucial to mitigate this issue.
- **Funding:** Securing funding for large-scale projects and missions is a constant challenge. Astronomers often rely on grants from government agencies, such as NASA and the National Science Foundation (NSF), and private foundations.
- **Data Management:** The increasing volume of astronomical data requires efficient storage, processing, and sharing solutions. Developing robust data infrastructure and collaboration platforms is essential.

Future Directions:

- **Next-Generation Telescopes:** Projects like the James Webb Space Telescope, the Extremely Large Telescope (ELT), and the Square Kilometre Array (SKA) promise to revolutionize our understanding of the universe by providing unprecedented observational capabilities.
- **Exoplanet Research:** Continued exploration of exoplanets, particularly those in the habitable zone, will focus on detecting biosignatures and understanding planetary atmospheres.
- **Gravitational Wave Astronomy:** The detection of gravitational waves by observatories like LIGO and Virgo has opened a new frontier in astronomy, allowing the study of cosmic events like black hole mergers and neutron star collisions.
- **Artificial Intelligence:** AI and machine learning will play an increasingly important role in analyzing vast datasets, automating observations, and even guiding telescope operations.

Astronomers' work is fundamental to advancing our knowledge of the cosmos. Their dedication to exploring the universe, combined with the continual development of new technologies and methods, ensures that our understanding of the vast expanse beyond our planet will continue to grow. As we push the boundaries of what is known, astronomers will remain at the forefront of discovering the secrets of the universe.

The Future of Human Space Exploration
The Future of Human Space Exploration

The horizon of human space exploration is brimming with promise, poised to be one of the most thrilling frontiers of scientific and technological progress. With ambitious plans for lunar colonies, human missions to Mars, significant contributions from private companies, and rigorous preparations for long-duration space travel, humanity stands on the cusp of becoming a multi-planetary species. Let's delve into these key aspects of future space exploration.

Plans for Lunar Colonies

The dream of establishing lunar colonies isn't just about landing on the Moon; it's about creating a sustainable human presence. Imagine a thriving outpost on the Moon, serving as a gateway for deeper space exploration.

Objective: The primary goal is to establish a permanent presence on the Moon, which will act as a stepping stone for missions to Mars and beyond.

Key Initiatives:

·Artemis Program: Led by NASA, the Artemis program aims to land the first woman and the next man on the Moon by 2024. This ambitious initiative is about more than just a return visit; it aims to create a sustainable lunar presence by the end of the decade.

- Lunar Gateway: Picture a space station orbiting the Moon, facilitating crew transfers, providing living quarters, and supporting scientific research. The Lunar Gateway will be the hub for lunar operations and deep space missions.
- Lunar Surface Missions: These missions will focus on exploring the lunar South Pole, believed to contain significant amounts of water ice. This ice is crucial for life support and can be converted into hydrogen and oxygen for rocket fuel.
-

·International Collaboration: The European Space Agency (ESA), Roscosmos (Russia), and JAXA (Japan) are all on board, collaborating on various aspects of lunar exploration. They are working together to build habitats, develop technology, and conduct joint missions.

Challenges and Solutions:

- Radiation Protection: The Moon lacks a protective atmosphere, exposing inhabitants to harmful cosmic rays and solar radiation. Innovative solutions include building habitats with thick walls, using regolith (moon soil) for shielding, and developing advanced radiation protection materials.
- Sustainable Living: Ensuring a continuous supply of food, water, and oxygen is critical. Closed-loop life support systems, lunar greenhouses, and water recycling systems are being developed to support long-term human presence.
- Energy Supply: Solar power will be the primary energy source, with solar panels deployed on the lunar surface. Research into nuclear fission reactors is also underway to provide reliable power during the long lunar nights.

Human Missions to Mars

Venturing further into space, human missions to Mars represent the next giant leap. Imagine astronauts exploring the red sands of Mars, building habitats, and searching for signs of past life.

Objective: The overarching goal is to explore and eventually colonize Mars, ensuring the survival of humanity by reducing our dependence on a single planet.

Key Initiatives:

·NASA's Mars Program: Building on the success of robotic missions like Curiosity, Perseverance, and the planned Mars Sample Return, NASA's Artemis program is laying the groundwork for human missions to Mars in the 2030s.

- o Mars Transfer Vehicles: Developing spacecraft capable of transporting humans to Mars is crucial. The Space Launch System (SLS) and the Orion spacecraft are key components of this effort.
- o Mars Habitat: Designing habitats that protect astronauts from radiation, provide life support, and sustain long-term living is essential. Concepts include inflatable modules, underground bases, and habitats built using Martian materials.

·SpaceX's Starship: SpaceX, led by Elon Musk, is developing the Starship spacecraft, designed for deep space missions, including Mars colonization. Starship aims to be fully reusable, significantly reducing the cost of space travel.

- o Launch and Landing: Starship will be launched atop the Super Heavy booster and land on Mars using its integrated landing system. The spacecraft is designed to carry up to 100 passengers, along with cargo and supplies.
- o Mars Colonization: SpaceX envisions establishing a self-sustaining colony on Mars, starting with small expeditions and gradually building a larger human presence.

Challenges and Solutions:

- • Radiation Exposure: During the journey to Mars and on its surface, astronauts will face high levels of radiation. Solutions include shielding spacecraft with water or hydrogen-rich materials and developing habitats with robust radiation protection.
- • Life Support Systems: Advanced life support systems that can recycle air, water, and waste are critical. Research into growing food on Mars using hydroponics or aeroponics is also underway.
- • Psychological and Physical Health: Long-duration space missions pose significant challenges to mental and physical health. Countermeasures include rigorous exercise regimens, psychological support, and researching the effects of microgravity on the human body.

The Role of Private Companies in Space Exploration

Private companies are revolutionizing space exploration, making it more accessible and sustainable. Imagine commercial spaceflights, suborbital tourism, and even orbital hotels.

Commercial Spaceflight:

- **SpaceX:** Founded by Elon Musk, SpaceX is a leader in commercial spaceflight. Its achievements include the Falcon 9 reusable rocket, Dragon spacecraft for cargo and crew missions, and the ambitious Starship project.
- **Blue Origin:** Founded by Jeff Bezos, Blue Origin focuses on reusable rockets and space tourism. Their New Shepard suborbital vehicle is designed for space tourism, while the New Glenn orbital rocket aims to support larger missions, including lunar landings.

Public-Private Partnerships:

- **NASA's Commercial Crew Program:** NASA collaborates with private companies to develop spacecraft capable of carrying astronauts to the International Space Station (ISS) and beyond. SpaceX's Crew Dragon and Boeing's Starliner are notable examples.
- **Lunar Landers and Gateway:** NASA's Artemis program includes contracts with companies like SpaceX, Blue Origin, and Dynetics to develop lunar landers. The Lunar Gateway will also involve contributions from commercial partners for various modules and logistics.

Space Tourism and Beyond:

- **Suborbital Flights:** Companies like Virgin Galactic and Blue Origin offer suborbital flights for space tourists, providing a few minutes of weightlessness and stunning views of Earth.
- **Orbital Hotels:** Future plans include developing orbital hotels for longer stays in space. Projects like Axiom Space's planned commercial module for the ISS and the proposed SpaceX Inspiration Mars mission aim to expand the space tourism market.

Preparing for Long-Duration Space Travel

Preparing for long-duration space travel involves overcoming numerous challenges. Imagine astronauts embarking on journeys lasting months or even years, venturing into the unknown reaches of space.

Life Support Systems:

- **Closed-Loop Systems:** Advanced life support systems that recycle air, water, and waste are essential for long-duration missions. NASA's Environmental Control and Life Support System (ECLSS) aboard the ISS is a prototype for such systems.
- **Food Production:** Growing food in space is critical for long missions. Research on hydroponics, aeroponics, and other soil-less farming techniques aims to provide astronauts with fresh produce.

Health and Safety:

- **Radiation Protection:** Innovative shielding technologies, such as magnetic fields, water-based shields, and advanced materials, are being developed to protect astronauts from cosmic radiation.

- **Counteracting Microgravity:** Prolonged exposure to microgravity can lead to muscle atrophy and bone loss. Exercise regimens, pharmaceuticals, and artificial gravity concepts (e.g., rotating spacecraft) are being researched to mitigate these effects.
- **Mental Health:** Long missions in confined spaces can impact mental health. Strategies include providing recreational activities, virtual reality environments, and psychological support.

Technology and Innovation:

- **Propulsion Systems:** Developing efficient propulsion systems, such as ion thrusters, nuclear thermal propulsion, and advanced chemical rockets, is crucial for reducing travel time and increasing mission feasibility.
- **In-Situ Resource Utilization (ISRU):** Utilizing local resources (e.g., extracting water from lunar or Martian soil) can reduce the need to transport supplies from Earth, making missions more sustainable.
- **Robotics and AI:** Robotics and artificial intelligence can assist with construction, maintenance, and scientific research on distant worlds, reducing the burden on human astronauts.

Training and Simulation:

- **Astronaut Training:** Comprehensive training programs prepare astronauts for the physical, technical, and psychological challenges of long-duration space missions. This includes simulations, underwater training, and survival training.
- **Simulated Missions:** Projects like NASA's HI-SEAS (Hawaii Space Exploration Analog and Simulation) and the Mars Society's Mars Desert Research Station simulate long-duration missions in isolated environments on Earth, helping researchers understand the challenges and develop solutions.

The future of human space exploration is bright and filled with potential. By addressing the challenges and leveraging the opportunities presented by new technologies and international cooperation, humanity is set to explore and potentially settle new worlds, ensuring our place as a spacefaring civilization. As we look to the stars, the dream of becoming a multi-planetary species becomes ever more tangible, heralding a new era of discovery and adventure.

Space and Society

As humanity looks up to the stars, the impact of space exploration reverberates across society, weaving into the fabric of our daily lives, shaping our future, and inspiring generations. Let's embark on a journey to understand how venturing into the cosmos influences our world and what it holds for us.

The Impact of Space Exploration on Society

Imagine a bustling city where people rely on GPS to navigate through the maze of streets. This technology, born from the need to explore space, now guides delivery trucks, emergency services, and everyday commuters. It's not just navigation; advancements in medical imaging, water purification systems, and communication networks all trace their origins to space exploration. The quest to understand the universe has driven the development of compact, efficient, and reliable systems, revolutionizing consumer electronics and robotics.

In this city, space exploration has fueled economic growth. High-tech jobs in engineering, research, and development thrive, supported by government and private investments. The space industry has become a beacon of innovation, creating new markets and stimulating economic activity. As we dream of mining asteroids for rare metals, the possibilities seem limitless, potentially transforming industries and economies on Earth.

Beyond the technological and economic impacts, space exploration fosters a sense of global unity. Nations come together, pooling their resources and expertise for missions that reach beyond our planet. This cooperation promotes peace and mutual understanding, transcending earthly borders. Meanwhile, children look up to the stars with wonder, their imaginations sparked by the possibilities of space. Educational programs flourish, inspiring the next generation of scientists and engineers who will continue to push the boundaries of human knowledge.

Viewing Earth from space, astronauts often experience a profound shift in perspective. They see our planet as a fragile, interconnected oasis in the vastness of space. This "Overview Effect" fosters a sense of responsibility towards our environment, urging us to address global challenges with the same determination that drives us to explore the cosmos.

Ethical Considerations in Space Exploration

But as we reach for the stars, we must tread carefully. Picture a pristine Martian landscape, untouched by human hands. Planetary protection policies ensure that our exploration does not contaminate these celestial bodies with Earth-originating microbes, preserving their integrity for scientific study. Similarly, strict protocols guard against the return of harmful organisms or materials to Earth from these distant worlds.

Space is not an infinite playground. The ever-growing cloud of space debris poses a significant risk to satellites and space missions. Imagine satellites orbiting the Earth, dodging debris from past missions. To ensure the safety and sustainability of space activities, we develop technologies to remove debris and adhere to guidelines that minimize further contributions to this cosmic junkyard.

The dream of mining asteroids and other celestial bodies brings forth questions of ownership and ethics. Who owns these resources, and how should the benefits be shared? The Outer Space Treaty provides some guidance, but new frameworks are needed as space resource utilization becomes a reality. It's crucial that these activities benefit all humanity and do not lead to environmental degradation or social inequity.

The well-being of astronauts is paramount. As they venture into the harsh environment of space, we must protect their rights and safety. This includes addressing the psychological and physical challenges of long-duration missions and ensuring fair treatment during and after their journeys. Inclusivity and diversity in space exploration are essential, opening opportunities for people of all backgrounds to contribute to and benefit from these endeavors.

The Role of Space Agencies and Policies

In the command centers of NASA, ESA, Roscosmos, and CNSA, plans are made for missions that will take humanity further than ever before. These space agencies lead significant missions, collaborating extensively to push the boundaries of our knowledge. The International Space Station (ISS) stands as a testament to international cooperation, where astronauts from around the world live and work together, conducting scientific research that benefits all.

The regulatory framework provided by the Outer Space Treaty ensures that space exploration is conducted peacefully and for the benefit of all humanity. As we look to the future, new policies will address the emerging field of space mining, managing space traffic, and ensuring the sustainability of space activities.

Inspiring the Next Generation

In classrooms and museums, children's eyes light up as they learn about space. Educational programs and partnerships with schools bring space science into the classroom, offering hands-on activities and experiments. Public outreach efforts, through media and interactive exhibits, demystify complex concepts and ignite a passion for discovery.
Role models like astronauts and scientists share their stories, inspiring young people to pursue careers in space and science. Mentorship programs connect students with professionals, providing guidance and support as they embark on their journeys into the unknown.

Competitions and challenges, like the International Science and Engineering Fair (ISEF) and NASA's Science Challenges, foster innovation and creativity. Students tackle real-world problems, designing solutions that could one day be used in space missions.

The Future Awaits

As we prepare for long-duration space travel, advanced life support systems, radiation protection, and health measures ensure the safety and well-being of astronauts. Research into efficient propulsion systems and in-situ resource utilization makes missions more sustainable. Robotics and artificial intelligence assist with tasks, reducing the burden on human astronauts.

The future of human space exploration is bright. By addressing ethical considerations and promoting inclusivity, we ensure that the benefits of space exploration are shared by all. As we push the boundaries of what is known, we inspire generations to come, fostering a spirit of curiosity and innovation that drives humanity towards the stars.

In the grand tapestry of the universe, our journey into space is just beginning. Each step we take into the cosmos brings new discoveries, challenges, and opportunities. Together, as one global community, we continue to explore, learn, and reach for the stars, forever expanding the horizons of human potential.

Chapter 5: The Enigmatic Quest for Alien Existence

The cosmos, a boundless theater of twinkling stars, enigmatic planets, and vast galaxies, hosts the compelling and age-old question: Are we alone in the universe? This chapter delves deep into humanity's relentless pursuit of extraterrestrial life, exploring the sophisticated technologies and scientific advancements propelling this quest. As we scan the heavens above, each celestial body from the tiniest moons to the sprawling, distant galaxies offers a potential clue in our search for life beyond Earth.

The journey toward discovering extraterrestrial existence is not just driven by modern technology but also by a rich history of astronomical achievements and philosophical inquiries that date back to ancient civilizations. These civilizations, looking up at the same stars we see today, pondered similar questions about our place in the universe. In the contemporary era, our tools have evolved from the naked eye to powerful telescopes orbiting Earth and roving robots on distant planetary surfaces. Projects like SETI (Search for Extraterrestrial Intelligence) and missions such as the Mars Rovers and the Voyager probes have extended our sensory reach far beyond our atmospheric boundaries, each contributing vital data that edge us closer to potential answers.

This chapter also highlights the groundbreaking detection of exoplanets that exist within the habitable zone of their stars—regions where conditions might be just right to support life as we know it. Advanced instruments like the Hubble Space Telescope, the Kepler Space Telescope, and the more recent James Webb Space Telescope have allowed us to peek into these distant worlds, analyzing their atmospheres and even searching for signs of biosignatures.

Additionally, we explore the implications of what discovering life beyond Earth would mean for humanity. From reshaping our scientific paradigms to challenging our philosophical and ethical frameworks, the discovery of alien life would be one of humanity's most profound revelations. The potential sociopolitical impacts, the philosophical implications about the nature of life, and the theological questions that would arise are all explored in detail.

As we stand on the brink of potentially transformative discoveries, this chapter not only chronicles our scientific endeavors but also reflects on the broader significance of finding life beyond our blue planet. It is a journey that tests the limits of our technology, challenges the breadth of our imagination, and might ultimately redefine humanity's role in the grand tapestry of the cosmos.

In the Shadow of the Cosmos: A Tale of Human Curiosity and Alien Conjecture

ILong ago, beneath the boundless star-strewn skies, our ancestors gazed up in wonder and trepidation. Each flickering light was a beacon from the beyond, each celestial event a storyline in the grand narrative written by the cosmos. As comets blazed across the firmament and meteors dashed through the night, ancient civilizations read these events as celestial scripts—messages or omens from realms unseen.

These early stargazers, whether cloaked in the drapery of Mesopotamia or the regalia of ancient China, were the first narrators in a saga that would span millennia. They linked the wanderings of celestial bodies to the fates of kings and empires, weaving the first threads of a connection that drew the heavens into the realm of the human experience.

As the wheel of time turned, bringing with it the intellectual awakenings of the Renaissance, the stage of the universe expanded. The celestial sphere was no longer a distant tableau adorned with the whims of gods, but a place governed by the same laws of motion and matter that ruled the Earth. This period, fueled by the minds of Copernicus, Galileo, and Kepler, transformed the celestial narrative from one of myth and superstition to one grounded in scientific inquiry.

Yet, the allure of the stars remained a fertile ground for the imagination. Literature burgeoned with otherworldly tales, reflecting humanity's perennial fascination with what might dwell in the vast unknown. H.G. Wells' The War of the Worlds brought invaders from Mars to the Victorian countryside, symbolizing the unsettling encroachment of the industrial age, and echoing deeper fears of the unknown and the other.

As the modern age dawned, the narrative of alien life shifted from the pages of fiction to the flickering frames of film. The Cold War era, rife with its own terrestrial fears of infiltration and obliteration, projected its anxieties into films like Invasion of the Body Snatchers, where the alien became a direct metaphor for the threat of ideological subversion.

Today, our engagement with the idea of extraterrestrial existence continues to evolve with our scientific advancements and cultural shifts. Films like Arrival challenge us to think of aliens not as invaders, but as messengers, urging a reevaluation of our place in the universe through the lens of communication and understanding rather than conflict.

In this chapter, we delve into how our perception of alien life has been sculpted by historical events and cultural narratives, exploring how these views reflect our deepest fears and highest hopes. It is a journey that spans from the primal awe of our ancestors to the sophisticated search signals we cast into the cosmos today, seeking not just answers but also understanding the grand narrative of which we are a part.

In the grand theater of the universe, Earth is but a single stage, and the search for extraterrestrial life extends the drama beyond our solitary spotlight into the dark, vast expanses of space. This pursuit, driven by an innate curiosity and a desire to answer one of humanity's most profound questions—"Are we alone?"—has become a symphony of scientific efforts spanning Earth and the cosmos.

The search is not a lonely quest confined to the whispers of distant stars but a dynamic exploration closer to our celestial home. NASA's SETI program, an iconic endeavor, extends our senses into space, using an array of giant radio telescopes to listen for whispers from distant civilizations. These telescopes scan the skies, searching for signals that break the monotony of cosmic radiation—a patterned beep or a deliberate pulse amidst the celestial static that could suggest the existence of advanced alien intellects.

Parallel to the silent wait for signals are the active explorations of robotic emissaries on other worlds. The Mars Perseverance Rover, a robotic geologist and astrobiologist, roams the red sands of Mars, scratching beneath its surface to unearth signs of past microbial life. Its every discovery is relayed back to Earth, where it is eagerly dissected for signs of organic material or chemical traces that could rewrite the narrative of life in the universe.

Beyond Mars, the ambitious Europa Clipper mission aims to soar through the icy plumes of Jupiter's moon Europa. This celestial body, shrouded in ice but possibly harboring an ocean beneath, represents the paradox of alien environments—hostile at first glance yet potentially a cradle for life. The mission seeks to capture and analyze the spray from these plumes, which may contain organic compounds or even microorganisms that thrive in the moon's subsurface ocean, much like the extremophiles on Earth.

The discovery of extremophiles—Earth's own protagonists in the saga of survival—has indeed stretched the boundaries of where life can exist. These hardy organisms thrive in boiling hot springs, acidic lakes, the frozen wastes of Antarctica, and even in the vacuum of space, enduring conditions that were once considered too extreme for life. Their existence has dramatically expanded the scope of habitable conditions and inspired a broader vision of what worlds might host life. From the sulfuric acid clouds of Venus to the methane lakes of Titan, life may not only exist but could be thriving in environments radically different from our own.

Each new discovery feeds back into the narrative of our search, suggesting that life might not only be possible in the varied tapestries of solar systems across the galaxy but might also be enduring, versatile, and unimaginably different from our own. In this cosmic quest, each hypothesis tested, each signal detected, and each molecule analyzed adds lines to the story of our universe—a story that we are only just beginning to understand.

The Drake Equation and Its Implications

Venture into the cosmos with the Drake Equation, a seminal formula proposed by astronomer Frank Drake in 1961. This equation doesn't just hypothesize about the existence of extraterrestrial civilizations; it frames our cosmic quest within the bounds of scientific inquiry, offering a method to estimate the number of civilizations in our galaxy with whom we might communicate. This chapter delves into each component of the equation, illuminating the scientific, technological, and philosophical underpinnings of our search for intelligent life beyond Earth.

Star Formation Rates: The Cosmic Crucible
Every journey through the cosmos begins with star formation. The stars are the forges of the universe, within which planetary systems are born. The rate at which new stars emerge in our galaxy sets the stage for the potential of habitable planets. Understanding these rates helps astronomers predict the abundance of star systems that might host life-bearing worlds.

Planetary Systems: Architectures of Complexity
The next step is to consider the fraction of stars that host planetary systems. Recent discoveries from telescopes and space missions like Kepler have dramatically increased our estimates, suggesting that planets are a common feature of many stars. This factor raises exciting possibilities about the variety and nature of these worlds.

Habitable Zones: Nurturing Life
Not all planets are created equal. The habitable zone, often referred to as the 'Goldilocks Zone,' describes the perfect range from a star where conditions are just right for liquid water to exist. This segment explores how scientists identify such zones and what this means for potential life.

Emergence and Evolution of Life
Life's emergence is perhaps the most mysterious aspect of the equation. We draw on knowledge of extremophiles and the earliest life forms on Earth to speculate on how life might arise in alien environments. Moreover, we consider the jump from simple life forms to complex, intelligent beings capable of altering their surroundings and developing technology.

Technological Communication: Reaching Out Across the Stars
Intelligence must be paired with the ability to communicate through interstellar space. This part of the chapter examines what technologies an advanced civilization might develop to send signals into the cosmos, and how we might detect such signals.

Longevity of Civilizations: The Great Unknown
The potential longevity of an alien civilization is crucial. Without sufficient time, even the most advanced civilizations might not survive long enough to be detected. We explore the factors that could influence the rise and fall of complex societies in the galaxy.

The Drake Equation does more than provide a numerical estimate—it invites us to ponder our place in the universe and propels our technological and exploratory efforts forward. As we refine each variable with better data and deeper understanding, our perception of the universe and our search for life beyond Earth evolve, turning science fiction into science fact.

Potential Forms of Alien Life: Beyond Earthly Boundaries

In the endless pursuit of understanding our universe, one of the most thrilling considerations is the potential variety of life beyond Earth: The imagination fuels science, and the diversity of alien life forms that might populate the cosmos provides a vivid tableau for scientific speculation. This section explores the theoretical foundations and creative conjectures about what forms extraterrestrial life might take, guided by the known laws of chemistry and physics.

Carbon-based Life: The Familiar Blueprint
Life on Earth is carbon-based; with carbon atoms forming the backbone of biological molecules. Carbon's ability to create complex and stable bonds makes it ideal for life processes. However, the prevalence of carbon-based life on Earth doesn't necessarily limit the possibilities elsewhere in the universe. Understanding carbon's role on Earth helps us appreciate why scientists consider it a likely candidate for life elsewhere but also underscores the potential for alternative biochemistries.

Silicon-based Life: A Theoretical Alternative
Silicon, like carbon, can form four bonds and is abundant in the universe, primarily as silicates and silicon dioxide. While no silicon-based life forms are known to exist on Earth, the element's chemical flexibility makes it a prime candidate for constructing alternative biochemical structures in environments unlike our own. This section delves into how silicon-based life might function, particularly on planets with high temperatures where silicon compounds could remain stable and play roles akin to those of carbon compounds on Earth.

Life in Alternative Solvents: Beyond Water

Water is the solvent of life on Earth, providing an environment where organic compounds can mix and react. However, other liquids, like ammonia or methane, which remain liquid at lower temperatures, could also serve as solvents for life on planets and moons where water is scarce. This part explores how ammonia-based life might exist in the subsurface oceans of icy moons like Saturn's Titan, where conditions could allow for chemical reactions different from those in water-based environments.

Non-organic Life: Venturing into the Unknown

The concept of non-organic life stretches the boundaries of traditional biology. This speculative realm includes forms of life that might not rely on molecular biology as we know it but could be based on entirely different principles, such as self-replicating patterns in plasma or other states of matter. Insights from the field of synthetic biology and artificial life provide a foundation for discussing how non-organic entities might evolve and sustain themselves.

Energy Sources and Life Processes

The potential energy sources for alien life are as varied as the environments they might inhabit. On Earth, life primarily relies on sunlight and chemical energy. In contrast, hypothetical alien life forms might harness other energy forms, such as radiation or geothermal energy, driving biochemistries that are unfamiliar to us. This section examines various energy acquisition methods that could support life, emphasizing the adaptability and creativity of life in meeting its fundamental energetic needs.

As our technology advances and our understanding deepens, the search for alien life continues to evolve from science fiction to a genuine scientific endeavor. Each hypothesis and model enriches our questions and directs our explorations, pushing the boundaries of known science. By contemplating the myriad possibilities for life beyond Earth, we not only expand our knowledge of the universe but also deepen our appreciation of life's potential complexity and resilience.

The prospect of making contact with extraterrestrial life is one of the most profound and contentious issues facing humanity. Beyond the immediate excitement and curiosity, the discovery of alien life has far-reaching implications that could alter our understanding of life, our place in the universe, and our interactions on a global scale.

Political Implications:

The confirmation of extraterrestrial existence could trigger a complex array of political actions. On the international stage, governments would need to determine how to collectively respond. The role of entities like the United Nations Office for Outer Space Affairs (UNOOSA), which currently oversees international space law and treaties, would become more pivotal. Protocols that currently exist, such as the Declaration of Principles on Activities Following the Detection of Extraterrestrial Intelligence, would be put to the test. This document advises signatory states to consult with humanity through the United Nations before responding to any alien communication.

Social Implications:

Socially, the impact would be unprecedented. The news could challenge many philosophical and religious frameworks that define modern civilization. Different societies may react with fear, wonder, or even denial. Historically, transformative discoveries have both united and divided societies—contact with extraterrestrial beings could have a similar polarizing effect depending on the nature of the discovery.

Psychological Implications:

On a personal level, individuals might experience a profound shift in their worldview. Psychologists predict a spectrum of responses, from existential crises and new philosophical movements to increased anxiety or profound inspiration drawn from the fact of not being alone in the universe. The psychological readiness of humanity to face such a reality is as significant as the technological capability to make contact.

Scientific and Technological Implications:

The discovery could accelerate technology development, particularly in space travel, communication technologies, and defensive capabilities, as humans seek to understand, communicate with, or even confront extraterrestrial entities. This technological leap could mirror the advancements seen during the space race of the mid-20th century.

Economic Implications:
The economic impacts could also be significant. New industries may arise with goals ranging from the commercialization of alien technologies to the development of space habitats and transportation systems designed for interstellar travel and trade. Conversely, the global economy could also face disruptions as traditional industries grapple with the new reality and the redirection of resources to space-oriented activities.

Ethical and Legal Implications:
The discovery of alien life would also ignite a complex debate over the ethical treatment of non-human life forms. Current space laws, which primarily focus on the exploration and use of outer space, might need overhauling to accommodate the rights and considerations of extraterrestrial beings. The development of a framework for peaceful and respectful interaction with any discovered entities would be a critical challenge.

Preparing for Contact

Realistically, international cooperation and comprehensive strategies are essential to prepare for such an event. Researchers and policymakers continue to advocate for the development of updated international protocols to manage the consequences of such discoveries effectively. Dialogue among global leaders, scientists, policymakers, and the public is crucial to cultivating a well-rounded approach to the potential discovery of extraterrestrial life.

The implications of contact with extraterrestrial life are vast and varied, touching upon virtually every aspect of human life. As we advance our capabilities in space and continue to search the cosmos, preparing ourselves not just technologically but also socially, politically, and psychologically for the possibility of extraterrestrial contact is imperative. This chapter aims not only to explore these multifaceted implications but also to encourage a thoughtful and informed approach to one of the most significant questions of our time: Are we truly alone?

First Contact: Implications and Impact

The very thought of first contact with extraterrestrial life conjures a mosaic of possible outcomes and scenarios. This section explores not just the direct effects of such a discovery but also the broader societal and geopolitical ramifications that may follow.

Societal Impact
Societal Impact

Philosophical and Religious Realignments: The confirmation of extraterrestrial life would trigger profound philosophical questioning and possibly a paradigm shift in many religious doctrines. Historically rooted perspectives might either adapt to incorporate the new reality or resist it, potentially leading to societal fragmentation or a renaissance of new thinking. Major world religions and philosophical schools would need to address the existential and ethical questions posed by the existence of life beyond Earth, affecting everything from worship to daily conduct.

Changes in International Relations and Security: The discovery could lead to unprecedented global cooperation, as nations might unite under a shared goal of understanding and interacting with extraterrestrial life. Alternatively, it might exacerbate international tensions as countries compete for control over extraterrestrial communications and technologies. Issues of national security could be redefined, considering the potential military implications of extraterrestrial technologies or the strategic positioning of Earth in a potentially broader inhabited cosmos.

Social Cohesion and Disruption: Depending on the nature of the contact, society could either pull together, driven by a common interest in the greater cosmos, or could fracture along new lines of those who embrace the new reality versus those who deny or fear it. Media portrayal and public perception of the extraterrestrial presence would play significant roles in shaping these outcomes.

The Potential for Collaboration or Conflict

Scenarios of Peaceful Exchange and Cooperation: Drawing from positive outcomes in human history, such as the peaceful exchange of ideas and mutual advancements during times of peace, we might hope for a scenario where humanity and extraterrestrial beings share knowledge and culture that enrich both parties. This could lead to technological and societal leaps forward, inspired by cooperative efforts in research, exploration, and cultural exchange.

Potential for Conflict and Competition: Conversely, human history is also riddled with conflict arising from fear, misunderstanding, or competition over resources. If extraterrestrial life forms have competing interests or if their needs intersect adversely with Earth's, the same tendencies could engender conflict. Scenarios might include competition over cosmic territories or resources, or even outright warfare, prompted by perceived threats or the strategic needs of survival in a two-species cosmos.

Interstellar Diplomacy and Warfare: The establishment of interstellar diplomacy would be crucial in navigating these complex relationships. Drawing on analogies from international relations on Earth, such as diplomatic protocols, treaties, and conflict resolutions, humanity could develop methods for peaceful interaction or conflict management with alien civilizations. However, if diplomatic efforts fail, humanity could find itself preparing for scenarios that seem lifted from science fiction, involving space-based warfare and defense strategies against technologically advanced beings.

Cultural Exchanges and Mutual Learning: In more optimistic scenarios, cultural exchanges could foster a profound mutual appreciation of diverse life forms, potentially leading to a new era of artistic and cultural renaissance influenced by alien perspectives and human traditions merging.

The possibilities of first contact range from enlightening exchanges to challenging confrontations. Preparing for all eventualities involves not only technological readiness but also a deep introspective look at human nature and society's capacity for adaptability and diplomacy. This section has explored the multifaceted potential impacts of such a pivotal event, urging a thoughtful, prepared approach to the unknown.

Unexplained Phenomena and UFO Sightings

The allure of the unknown has always drawn human attention skyward, where the unexplained phenomena and unidentified flying objects (UFOs) evoke a mixture of curiosity, skepticism, and fear. This section delves into some of the most compelling UFO sightings, their investigations, and the ensuing governmental responses that have both illuminated and obscured our understanding of these enigmatic occurrences.

Overview of Famous UFO Sightings

Roswell Incident (1947): The Roswell Incident remains the epitome of UFO folklore. After a rancher discovered unidentifiable debris on his property near Roswell, New Mexico, the U.S. military initially reported capturing a "flying disc." However, this statement was quickly retracted, claiming the debris was from a crashed weather balloon. Over the decades, the incident has been shrouded in conspiracy theories regarding alien spacecraft and government cover-ups, continually fuelling public intrigue and skepticism.

Phoenix Lights (1997): This event involved the sighting of mysterious lights over Phoenix, Arizona, observed by thousands. Unlike any conventional aircraft, these lights formed a large V-shaped pattern that moved across the sky and were visible for several hours before disappearing. Despite official explanations of military flares, the synchronized movement and prolonged visibility have led many to continue questioning the true nature of the lights.

Tic Tac UFO Encounter (2004): In one of the most credible UFO sightings, fighter pilots from the USS Nimitz encountered an unidentified "Tic Tac"-shaped object off the coast of Southern California. Captured on gun-camera footage, the object exhibited extraordinary aerodynamic capabilities that defy known aviation technology, including incredible acceleration and maneuverability. This encounter, officially released by the Pentagon, has prompted calls for serious scientific investigation into unidentified aerial phenomena.

In recent years, the curtain over the secretive world of unidentified flying objects has been partially lifted, signaling a new era of transparency. Notably, the Pentagon's move to declassify several videos depicting encounters with UFOs marks a pivotal shift. This openness aims to destigmatize UFO sightings among military personnel, encouraging them to report their observations without fear of ridicule or career repercussions.

The establishment of the Unidentified Aerial Phenomena (UAP) Task Force underscores this evolution. It acknowledges the potential security threats these unidentified objects could pose, bringing a topic once relegated to the fringes of conspiracy theories into the spotlight of legitimate national security deliberations. The task force's reports and findings have begun to sketch a more nuanced picture of the phenomena, suggesting a complexity and behavioral pattern that challenges conventional aerospace technology understandings.

Credibility and Public Reaction

The credibility of UFO sightings has significantly shifted, supported by the caliber of the witnesses—often highly trained pilots and military personnel—whose expertise and reliability enhance the veracity of their reports. This credibility is further bolstered by advancements in technology that allow for the capture of clearer, more definitive visual data than ever before. High-resolution cockpit displays, infrared systems, and radar technologies have all captured phenomena that defy easy explanation or conventional classification.

Public interest in UFOs has surged alongside these developments, moving the conversation from the realms of science fiction to a more scientifically respected enigma. This paradigm shift is evident across various platforms—from the surge in academic papers and symposiums discussing the phenomena to the portrayal of UFOs in movies, books, and television, reflecting a cultural readiness to accept and understand the unknown.

The response from the public and the media illustrates a growing consensus that what was once dismissed as imaginative fantasy may be grounded in reality. Discussions are no longer about proving the existence of UFOs but rather understanding their behavior, origin, and technology. Policy makers, scientists, and governments are increasingly engaged in dialogues about how to respond to and engage with phenomena that could potentially reshape our understanding of physics and the universe.

Broader Implications and Philosophical Impact
Broader Implications and Philosophical Impact

As we delve deeper into the mystery of UFOs and other unexplained phenomena, the potential for significant philosophical and societal shifts looms large. The acknowledgment of extraterrestrial life, whether microbial or intelligent, could have profound implications for every aspect of human culture and knowledge. It challenges our understanding of life, physics, and even spirituality, offering new perspectives on our place in the universe.

The exploration of these phenomena not only bridges the gap between skepticism and belief but also enriches the scientific community's quest for truth. Whether or not contact is made in our lifetime, the very act of searching enhances our understanding of the cosmos, pushing the boundaries of human technology and thought. In contemplating the existence of otherworldly visitors, we are invited to reflect on our own civilization, its potential legacy, and how we might communicate with species who may share nothing in common with us but the cosmos itself.

Ultimately, the pursuit of these mysteries encourages a broader, more inclusive view of our place within the grand tapestry of the cosmos. It prompts us to question not merely the existence of alien life but also the extensive implications of such a discovery on our global society, technologies, and philosophies.

Chapter 6: Cosmic Curiosities: Astonishing Astronomy Facts

In the vast canvas of the cosmos, where stars are born from cosmic dust and black holes whisper secrets from the dawn of time, every corner of the universe brims with wonders beyond imagination. This chapter delves into the peculiarities of celestial bodies and phenomena that seem more like science fiction than fact, offering a glimpse into the extraordinary nature of our universe. Prepare to have your mind expanded by the astonishing realities that space has to offer.

Stellar Surprises

Fact 1: The Cosmic Lifecycle of Stars Stars are not just points of light scattered across the night sky; they are dynamic and vibrant entities undergoing dramatic transformations. A star's life begins in a nebula, where vast clouds of gas and dust collapse under their own gravitational pull, initiating nuclear fusion. This process lights up the star for millions to billions of years, depending on its size. The finale of a star's life can be equally spectacular—if massive enough, a star ends its life in a supernova, an explosion so bright that it can outshine entire galaxies and spread the elements necessary for life into the cosmos.

Fact 2: Pulsars and Magnetars: The Lighthouses and Powerhouses of the Galaxy Pulsars and magnetars present some of the universe's most extreme conditions. Pulsars, the rapidly spinning remnants of supernova explosions, emit beams of electromagnetic radiation that can be observed as pulses when they sweep past Earth—much like cosmic lighthouses. Magnetars are a rare type of neutron star with the most powerful magnetic fields known in the universe. These magnetic fields are so intense that they could erase credit cards from hundreds of thousands of miles away and have even been known to disrupt satellites in orbit.

Planetary Oddities

Fact 1: Exoplanetary Extremes Beyond our solar system, planets defy the familiar conditions found on Earth. Take HD 189733b, a planet where astronomers have detected glass rain—silicate particles in its atmosphere condense into glass that whips sideways in ferocious winds exceeding 5,400 mph (about 7 times the speed of sound). Another bizarre world, 55 Cancri e, is thought to have a surface composed largely of diamond, a result of its carbon-rich composition and the immense pressures within its interior.

Fact 2: Wonders of Our Solar System Even within our solar system, the planets and moons present landscapes and weather that challenge Earthly norms. Jupiter's Great Red Spot, a storm larger than Earth, has raged for centuries, offering a spectacle of persistent atmospheric turmoil. Venus, cloaked in a thick atmosphere of carbon dioxide, experiences surface pressures so intense and temperatures so high that lead would melt into a puddle, creating a surface that is anything but hospitable.

Each fact not only underscores the diversity and the sheer scale of phenomena within our universe but also enhances our appreciation for the complexities of celestial mechanics and compositions. These insights not only broaden our knowledge but also fuel our curiosity about what other secrets the universe holds, waiting to be discovered.

Galactic Wonders

Fact 1: The Vastness of Galaxies and Secrets of the Milky Way
In the immense tapestry of the universe, galaxies are the grand archives of cosmic history, each a whirlpool of ancient light, time-bound energy, and celestial matter. The Milky Way, our own galactic home, is a sprawling barred spiral galaxy, with elegantly swirling arms that stretch some 100,000 light-years from end to end. Within this galaxy, billions of stars including our Sun, perform an age-old dance around a common center.

Hidden from the naked eye, shrouded behind dense cosmic dust, lies Sagittarius A*, a supermassive black hole whose mass is roughly four million times that of our Sun. This colossal entity doesn't just anchor the galactic ballet with its immense gravitational pull; it shapes the fate of stars and the structure of our galaxy. As stars orbit this massive object, some come perilously close, leading to spectacular cosmic phenomena, such as high-energy flares and dramatic shifts in their trajectories, which astrophysicists eagerly study to better understand the dynamic processes driving galactic evolution.

Fact 2: The Diversity and Mysteries of Other Galaxies
Venturing beyond the Milky Way, the universe presents a gallery of galactic diversity that defies imagination. Each galaxy, from spirals and ellipticals to irregulars, tells a unique story of cosmic fortune and survival. Take, for example, NGC 454, an interacting galaxy pair located millions of light-years away. This system, often called the "Cannibal Galaxy," provides a stark visual of galactic interaction, where a larger elliptical galaxy is slowly consuming its smaller companion.

This vivid demonstration of galactic cannibalism underscores the dynamic and sometimes violent nature of the universe.

In another part of the cosmos, NGC 1052-DF2 challenges our very understanding of galactic makeup. This ultra-diffuse galaxy appears to contain no dark matter, a finding that contradicts prevailing theories that dark matter is a fundamental component of galaxies and essential for their formation. Such anomalies compel astronomers to rethink existing models and explore new theories that might explain these cosmic puzzles.

Mysteries of the Cosmos

Fact 1: The Enigmatic Nature of Dark Matter and Dark Energy

The universe is a mystery of shadow and substance, with the vast majority of its mass-energy content hidden in forms that escape direct detection. Dark matter, though invisible, reveals its presence through the gravitational forces it exerts on galaxies and galaxy clusters. It acts as the cosmic glue that holds these structures together, yet what it is made of remains one of the great questions in modern physics.

Equally mysterious is dark energy, a perplexing force that is driving the accelerating expansion of the universe. This mysterious energy, making up about 68% of the universe's energy content, is pushing galaxies apart at an ever-increasing speed. Its nature is perhaps the most profound mystery in the cosmos, representing a frontier in astrophysics that challenges our understanding of the universe's ultimate fate.

Fact 2: The Accelerating Expansion of the Universe

The discovery that the universe is not just expanding, but accelerating, was a watershed moment in astronomy. This revelation came from observations of distant supernovae, which appeared dimmer than expected, suggesting that they were farther away than their redshifts implied. This acceleration points to the influence of dark energy, a repulsive force that pervades all of space.

As we continue to probe these celestial enigmas with more sophisticated telescopes and sensors, each discovery peels back a layer of the universal mysteries, inviting us to rethink our place in this vast cosmos. The journey through these galactic wonders and cosmic mysteries does more than satiate our curiosity; it propels our scientific theories into new realms, forever altering our philosophical and existential frameworks.

The End

As we conclude this exploratory odyssey across the universe, we are reminded of the vastness that surrounds us—a cosmos brimming with mysteries that have both baffled and inspired generations of astronomers, philosophers, and dreamers alike. This book has been a guide through the celestial phenomena and the enigmatic realms that define our universe, from the swirling galaxies in the distant cosmos to the peculiarities of planets both within and beyond our solar system. Throughout these pages, we've seen how stars are born, live, and die in spectacular supernovae, seeding the cosmos with the elements that make up worlds and potentially, other forms of life. We've ventured into the heart of galaxies, including our own Milky Way, to confront the supermassive black holes that lurk at their centers, governing their majestic spirals yet reminding us of the powerful forces at play in the universe.

Our journey has also taken us through the theoretical landscapes of modern astrophysics, where the dark mysteries of matter and energy challenge our understanding of the physical laws that govern the cosmos. We've speculated on the existence of other universes, each possibly with its own laws of physics and unique constellations, expanding our conception of reality to scales beyond our wildest imaginations.

The search for extraterrestrial life has opened new discussions on the potential for contact and the profound implications such an event would hold for humanity. It has reshaped our philosophical, ethical, and, potentially, spiritual understandings of our place in the universe. Whether through the decoding of a signal from a distant star system or the discovery of microbial life on a moon in our solar system, each step brings us closer to answering the age-old question: Are we alone in the universe?

As we gaze into the night sky, the stars look back at us with the light of billions of years past, each one a reminder of the enduring quest for knowledge that drives us. The universe is a grand canvas, and our understanding of it is still in its infancy. Each discovery adds a brushstroke to the masterpiece that is our ever-evolving cosmic map.

Let this book be a testament to human curiosity and a beacon for future explorers of the cosmos. The path to understanding the universe is infinite, filled with new technologies, future missions, and, undoubtedly, surprises that will continue to challenge and expand our understanding.

In closing, we are left with a sense of humble awe and a renewed commitment to explore, to learn, and to connect with the universe around us. The cosmos is vast, and its wonders are waiting to be discovered. As you turn the final page, may your mind wander to the infinite possibilities that await and may your heart be ever curious about the wonders beyond our world. The journey does not end here; it is just another beginning in our endless quest to understand the cosmos and our place within it.

Embarking on Your Own Observations

As we embrace the boundless journey of cosmic exploration highlighted in this book, the next section offers a practical guide to beginning your own celestial observations. Here, we will introduce a selection of telescopes, carefully chosen to empower amateurs and seasoned stargazers alike. These recommendations are crafted to help you navigate the night sky with clarity and awe, transforming your curiosity into a tangible experience of the universe's grandeur.

Best Telescopes for Amateur Astronomers

Embark on the fascinating journey of amateur astronomy, where the cosmos unfolds before your eyes. Stargazing can be a transformative experience, connecting us with the vast universe and offering a tangible sense of wonder. Selecting the right telescope is your first step toward unlocking the mysteries of the night sky.

Types of Telescopes

Understanding the basic types of telescopes is crucial in making an informed decision:

- Refractors: Known for their long, slender tubes, refractor telescopes use lenses to bend light and form an image. They are excellent for viewing the moon and planets and offer high contrast and sharp images. They're typically low maintenance but can get expensive with increasing aperture size.
- Reflectors: These telescopes use mirrors instead of lenses, which allows them to be larger and more affordable. They're ideal for viewing faint objects like distant galaxies and nebulae. However, their open tube design may require regular adjustments and cleaning.
- Compound (Catadioptric): Combining features of both refractors and reflectors, these telescopes are versatile and portable. They use both lenses and mirrors to provide detailed images and are excellent for both celestial and terrestrial viewing. They often come with advanced features like computerized controls.

Key Features to Consider

Choosing a telescope is about more than just the type; here are the features to consider:

- Aperture Size: The diameter of the telescope's primary optical component. Larger apertures gather more light, which is essential for seeing faint objects in deep space.
- Focal Length: A longer focal length increases the magnification and sharpness of the image but may require a larger, less portable setup.
- Mount Type: Alt-azimuth mounts are simpler and good for beginners, while equatorial mounts are better for tracking the sky's motion, ideal for long exposure astrophotography.

Recommended Telescopes for Beginners

For those starting their stargazing journey, consider these models:

- Celestron NexStar 5 SE Telescope: A user-friendly compound telescope that combines portability with advanced features like a computerized object database.
- Orion SkyQuest XT8 Classic Dobsonian Telescope: This reflector offers a large aperture at an affordable price, excellent for viewing deep-sky objects.
- Meade Instruments Infinity 102mm AZ Refractor Telescope: An excellent choice for beginners, this refractor comes with everything needed to get started and is particularly good for viewing planets and the moon.

Accessories and Enhancements

Enhance your viewing experience with these essential and advanced accessories:

- Eyepieces: Invest in a variety of eyepieces to change magnification and field of view.
- Filters: Use filters to enhance the visibility of planetary features and reduce light pollution.
- Cameras: Consider a dedicated astronomy camera or a DSLR adapter to start capturing stunning images of the cosmos.

Tips for Successful Stargazing

Maximize your stargazing experience with these tips:

- Location: Find a dark spot away from city lights to improve visibility of faint objects.
- Celestial Events: Plan your sessions around meteor showers, planet alignments, or other celestial events for a more rewarding experience.

Exploring the night sky with a telescope is an enriching activity that opens up new perspectives on the universe and our place within it. With the right equipment and knowledge, the wonders of the cosmos are just a lens away. Remember, every clear night is an opportunity to see beyond the everyday world and into the depths of space.

Captured through the lens of a telescope, these images present a genuine glimpse into the celestial wonders of the universe.

Captured through the lens of a telescope,
these images present a genuine glimpse
into the celestial wonders of the universe.

Captured through the lens of a telescope, these images present a genuine glimpse into the celestial wonders of the universe.

Captured through the lens of a telescope, these images present a genuine glimpse into the celestial wonders of the universe.